Immobilized Enzymes in Food and Microbial Processes

Immobilized Enzymes
in Food and Microbial Processes

Edited by

Alfred C. Olson

Western Regional Research Laboratory
Agricultural Research Service
U.S. Department of Agriculture
Berkeley, California

and

Charles L. Cooney

Department of Nutrition and Food Science
Massachusetts Institute of Technology
Cambridge, Massachusetts

PLENUM PRESS • NEW YORK AND LONDON

Library of Congress Cataloging in Publication Data

Main entry under title:

Immobilized enzymes in food and microbial processes.

"Proceedings of a symposium on immobilized enzymes held at the 166th national meeting of the American Chemical Society, Chicago, Illinois, 1973."
Includes bibliographical references.
1. Immobilized enzymes—Industrial applications— Congresses. I. Olson, Alfred C., ed. II. Cooney, Charles L., 1944-   ed. III. American Chemical Society. [DNLM: 1. Enzymes—Congresses. 2. Food additives—Congresses. 3. Food—Processing industry —Congresses. QU135 I33]
TP248.E5I47   661'.894   74-9866
ISBN 0-306-30789-8

Proceedings of a Symposium on Immobilized Enzymes
held at the 166th National Meeting of the
American Chemical Society, Chicago, Illinois, 1973

© 1974 Plenum Press, New York
A Division of Plenum Publishing Corporation
227 West 17th Street, New York, N.Y. 10011

United Kingdom edition published by Plenum Press, London
A Division of Plenum Publishing Company, Ltd.
4a Lower John Street, London W1R 3PD, England

Copyright is not claimed in any portion of this work written by a United States Government Employee as a part of his official duties. Reference to a company or product name does not imply approval or recommendation of the product by the U. S. Department of Agriculture to the exclusion of others that may be suitable.

Printed in the United States of America

Preface

In the last five years the potential value of immobilized enzymes has captured the imagination of an increasing number of scientists and engineers. The concept of being able to create an immobilized derivative of an enzyme which has long-term stability and is able to be recovered and reused is fascinating, to say the least. Since the industrial application of enzymes has been mostly in the food and microbial process industry it is not surprising that many of the applications of immobilized enzymes considered for commercial development fall within the area of this industry. It is for this reason that we organized a symposium on immobilized enzymes for the 166th National Meeting of the American Chemical Society. Appropriately, the symposium was jointly sponsored by the Division of Agricultural and Food Chemistry and the Division of Microbial Chemistry and Technology. Although there were at least half a dozen symposia at other meetings on various aspects of immobilized enzyme technology in the preceding ten months none had specifically addressed themselves to food and microbial processes and none had been held at a meeting such as the National ACS gathering, which is not only large but multidisciplinary. The enthusiastic response to this symposium prompted us, at the invitation of Plenum Press, to publish the proceedings of this symposium.

This volume is by no means a complete treatise on immobilized enzymes in the food and microbial process industry. It does, however, pinpoint some of the major areas of research, the types of approaches taken and the state-of-the-art with regard to these approaches. Because research on immobilized enzymes is an active and dynamic field, we felt that a volume such as this would have maximum value only if it was published as soon as possible after the symposium.

We attempted to balance the tone of the meeting by inviting speakers from universities, industry and government laboratories. In addition, for this volume, we have included a contribution by R. A. Messing, which was presented at another session of the Division of Agricultural and Food Chemistry, and a contribution by B. K. Hamilton, C. K. Colton and C. L. Cooney.

v

The introduction to this volume, written by Dr. E. K. Pye, outlines the present and future trends for enzyme technology. The next three papers discuss the application of immobilized enzymes to milk and lactose processing. Drs. T. Richardson and N. F. Olson have considered a variety of enzyme applications to milk processing and the papers by Dr. J. H. Woychik and his co-workers and Drs. A. C. Olson and W. L. Stanley examine more specifically the use of immobilized β-galactosidase (lactase). In the next chapter Dr. J. Reynolds discusses the application of α-galactosidase in a continuous flow reactor to the hydrolysis of oligosaccharides in soy milk and beet sugar molasses.

One of the newest industrial scale applications of immobilized enzymes is the production of glucose/fructose syrups from glucose utilizing immobilized glucose isomerase. In this manner Clinton Corn Products at Clinton, Iowa is making 100,000 pounds annually of this syrup. Because of its relevance to industrial scale processing, glucose isomerase has become quite an active focal point for research. M. J. Kolarik and his co-workers at Purdue University present some of their work in Chapter 6, and B. K. Hamilton, C. K. Colton and C. L. Cooney discuss the overall aspects of glucose isomerase as a case study in enzyme technology. The last paper concerned with process development was prepared by Drs. K. L. Smiley and co-workers on degradation of starch waste. In the next three chapters, consideration is given to several approaches to techniques for immobilization. This is followed by a theoretical treatment of the effect of diffusional limitation on the behavior of immobilized enzymes.

Having examined some aspects of enzyme processes, as well as techniques for immobilization, a hypothetical case, the use of enzymes in the synthesis of carbohydrates is considered. This paper was presented by Dr. J. Shapiro at the ACS Meeting. This chapter, prepared in collaboration with Drs. J. Adams and J. Billingham is presented here with the sad acknowledgment of Dr. Shapiro's untimely death shortly after the ACS Meeting.

The concluding chapter in the book was prepared by Dr. George Tsao who has guided the enzyme technology program of the National Science Foundation's RANN program for the past year. In this paper, Dr. Tsao proposes a long-term approach to biotechnology which appropriately would take off from our accumulated experience and knowledge of enzyme technology.

We are indebted to all of the authors for their preparation of these manuscripts and their cooperation in meeting a tight publication schedule. We apologize for any errors or omissions in the text that escaped our scrutiny. Appreciation is due Dr. Daniel I. C. Wang from the Division of Microbial Chemistry and Technology and Dr. S. J. Kazeniac from the Division of Agricultural and Food Chemistry for their help in the organization of the symposium.

Charles L. Cooney
Alfred C. Olson

Contents

PRESENT AND FUTURE TRENDS IN ENZYME TECHNOLOGY

AND ITS APPLICATION

E. Kendall Pye

Department of Biochemistry
University of Pennsylvania
Philadelphia, Pa.

To the casual observer, the bulk enzyme industry does not
appear to be a particularly significant factor in the U.S. and
world economy. Since its initial flourishings in the early part of
this century it has not been subject to the spectacular growth
which has projected other science-based industries, such as elect-
ronics and petrochemicals, into their present roles of major economic importance. The reasons for this relatively mediocre performance are complex, but several can be suggested here. Among these
is the fact that for practically its entire history the industry
has suffered from a lack of basic scientific understanding of its
complex biological products and production processes. Only over
the last decade or so have significant amounts of this knowledge
become available, thus providing the opportunity for a rational
rather than an empirical development of enzyme production and applications.

Another possible reason for the relatively poor growth of the
bulk enzyme industry is that in general it does not cater to a large
primary consumer market but instead sells enzymes almost exclusively
to other industries as processing materials. Consequently the growth
of the market for enzymes has been determined,by and large,by the
growth in the market for the ultimate products and by new applications. More recently, strong government regulations,required to
protect human health and safety, have tended to inhibit the development and rapid introduction of potential enzyme applications in
many sectors of our economy.

These factors might well be among the major reasons why the
bulk enzyme industry during its lifetime has developed only a
relatively narrow range of products and applications despite the

extensive array of enzymes available in nature. The traditional products of the industry, the hydrolytic enzymes, are still its major source of revenue, probably because these enzymes are reasonably stable and free from the handling and storage problems of other classes of enzymes and therefore find easy application in other industries. Nevertheless, despite the dramatic advances which have occurred since the 1950s in the basic biological sciences and in enzymology itself, the industry has developed only little in the way of novel large-scale applications, and, like the brewing industry with which it shares several common aspects, until recently it has been slow to change from its traditional methods and markets. One possible reason for this lack of significant innovative advances in enzyme application and technology at a time when the basic sciences were mushrooming, is that for a long time now the overwhelming majority of biological scientists, and the national research efforts in the biological sciences in general, have been directed towards basic medical problems and health-related applications. In this euphoric flurry of biological research, much of it maintained by large government funding, the application of enzymes to industrial processes received very little attention from the basic scientists and the development of new commercial applications for enzymes was left primarily in the hands of the industry itself and those already narrowly concerned with other industries, such as food, leather, brewing and textiles.

In the light of all this it is unfortunate that on the one recent occasion when the bulk enzyme industry had an opportunity for spectacular growth into a large consumer product, its market was panicked into a precipitous and senseless demise. I am, of course, referring to the use of proteases in laundry products. This episode, which was especially catastrophic for the newer, smaller companies having no traditional markets on which to fall back, might well have a long-term, insidious effect on the morale and the research and development efforts of the bulk enzyme industry.

The statistics of this episode are quite revealing. In Europe, where not very popular enzyme-containing laundry soaking agents have been on the market for over 50 years, the enzyme industry was able to develop enzyme preparations having much greater resistance to the non-physiological conditions prevailing in suds. This sparked the development from enzymatic presoaks to effective enzyme-containing washing powders. A revolutionary expansion of enzymatic laundry products then occurred, especially in Holland where the share of such products in the total washing agent market rose from 10 per cent in 1964 to 50 per cent in 1970. In the U.S. the market was developed later but even more rapidly. In 1967 the enzyme-containing laundry product was only 5 per cent of the total U.S. market, but in two years this figure had risen to 50 per cent. Of course, a rapid expansion of protease production capacity was necessary in order to provide the many thousands of tons required annually. When adverse publicity in the period 1970-1971 cut the market in the U.S. drastically some of the smaller companies were forced out of business. Although the enzymatic proportion of the market in the U.S. is now, in 1973, quoted as being back up in the range of 25 per cent, the market is apparently supplied almost entirely by the larger companies.

In considering these points it is a reflection on the enzyme industry and on the technology that while a patent for the use of enzymes in the laundry area was issued in 1913 to Otto Röhm (1), it then took over half a century before the necessary development took place for this market to be exploited to any significant degree. Another point to ponder is that despite the industries tradition and experience in enzyme production and handling, the commercialization of purified enzymes for analysis and research, which has shown a very significant growth since the 1950s, has not in general been carried out by the bulk enzyme industry. Instead, this market has largely been exploited by newer, smaller companies, several of which were formed by previously academically-based individuals who saw the needs and the opportunities which existed.

CURRENT TRENDS

In our consideration of the present and future trends of enzyme technology and its application we must start from this current view of the enzyme industry, its markets as they exist today, and the state of the technology which underlies it. The bulk enzyme industry still relies almost entirely on the production and sale of tonnage quantities of relatively simple enzymes, primarily to the food and other consumer industries. The major products of the bulk enzyme industry have a number of features in common and may be regarded as "first generation" enzymes in terms of their production technology and mode of application. As shown in Table I most bulk enzymes now produced and used in industry are hydrolases (amylases, proteases, lipases, pectinases, cellulases, etc.); they are used almost exclusively in a soluble, non-recoverable form; they are mostly extracellular enzymes obtained from deeptank and surface fermentations of microorganisms and can be viewed as the digestive enzymes of microorganisms; few, if any, have a cofactor requirement; they are sold with little or no purification, usually as dried powders, and they are of low cost.

These enzymes are used mostly in traditional fashions as additives or processing aids in the baking, meat, dairy, canning, fruit juice and other industries. However, one present trend is toward the appearance of some newer applications for first generation enzymes. One example is in the treatment of commercial soft woods, such as Norway Spruce. These trees, when freshly felled, are resistant to treatment with chemical preservatives, even when these are

TABLE I

Common Features of Most Bulk Enzymes Presently
used in Industry   - First Generation Enzymes

(a) Hydrolases (amylases, proteases, lipases, pectinases,
        cellulases, etc.)

(b) Used in soluble, non-recoverable form.

(c) Extracellular enzymes obtained primarily from microbial
        fermentations (exceptions, e.g. pepsin, papain).

(d) Few, if any, have a cofactor requirement.

(e) Marketed with little or no purification.

(f) Low cost.

applied under pressure. Pretreatment with pectolytic enzymes in
enclosed tanks has been shown to increase the permeability of the
wood and to allow more effective penetration of preservatives (2).
Should it be shown that this application has significant value and
is economically sound, a dramatic increase in the market for pectinases could well result. In addition to this, other novel applications for the first generation enzymes are now appearing, especially
in the food industry. As an example, at least one Japanese company
(3) now markets an enzyme preparation containing naringinase activity. Naringin, which is widely found in fruits and juices, has a
bitter taste and is hydrolyzed by naringinase to form prunin. A
flavonoid glucosidase, also present in the enzyme preparation, then
hydrolyzes the prunin into non-bitter naringenin. The same company
markets a hesperidinase-containing preparation which hydrolyzes
hesperidine, a white turbidity found in the syrups of canned man-
darin oranges, to the non-turbid material hesperetin (3). Also,
anthocyanase preparations are marketed which find a use in the
decolorization of grape, peach, strawberry and other fruit products
(3). Other recent applications for first generation enzymes include
the use of melibiase to hydrolyze the small amount of raffinose
which impairs the crystallization of beet sugar, causing a lowering
of quality and yield of the final product, sucrose (4). The utilization of this enzyme might result in significant economies and increased yields in the worlds beet sugar production which is in
the range of 9 million tons in the USSR, 8 million tons in Europe
and 2.5 million tons in the U.S.

Another first generation enzyme which is rapidly finding commercial application is cellulase. This enzyme is finding increasing
use as an additive in animal feeds to increase cellular digestion

and greater utilization by the animal of the nutritive value of vegetable feeds. It is also considered to be of value in increasing the recovery of starch from starch pulp (5). In the near future it may also play a large and highly significant role in the production of glucose syrups from cellulosic wastes and paper (6).

Several of these novel applications of the newer first generation enzymes have only limited market potential but others, such as the wood treatment with pectinases, the melibiase treatment of beet sugar and the production of glucose from cellulose using cellulases may well have major economic impacts. In addition the sale of amylases and β-glucanases could also receive a large boost if their use becomes as generally accepted in malting and brewing processes in the U.S. as it is in other parts of the world (7).

Microbial rennin is another enzyme preparation which shows considerable promise. With animal rennin being in short supply world-wide, microbial rennins appear to have a strong future in the production of hard cheeses, although some development still appears to be necessary in order to make their relative milk clotting and proteolytic activities completely comparable to those of animal rennin. While several microbial rennins are presently in use any new preparations must obtain FDA clearance before use in the U.S. If they could replace animal rennins totally, they have a potential annual world market value of approximately $23 million (8).

There are many other recent applications of first generation enzymes which are either being tested or are now undergoing considerable expansion, but which unfortunately can only receive a superficial mention here. Included among these are the use of isoamylase for the production of maltose from starch, invertase for the production of invert sugar from sucrose, lactase for the removal of lactose from whey, proteases for the upgrading of meat, keratinase for leather processing and modification of wool and hair, dextranase for the removal of tooth plaque, tannase for the removal of tannic acid in foods, and penicillin amidase which is used in the production of 6-APA from penicillin G.

Taking all of this information together, it is clear that despite the relatively slow progress of the enzyme industry up until the last decade, there is now a very marked trend towards a broader application of the traditionally produced enzymes and a sharp expansion in the search for applications for more novel hydrolytic enzymes. Many of these applications could have a significant economic impact and it might be expected that their production and sales will increase accordingly over the next decade. Of course, much depends on government and consumer acceptance. For example, it is estimated that the total retail market in the U.S. for enzymatic digestive aids is in the range of $12 million annually, provided there was the necessary government and consumer acceptance (8).

While the potential markets over the next decade for many of these more novel enzyme applications cannot be accurately assessed at this time, assessments have been made of the market potentials of various classes of enzymes on the basis of present and clearly foreseeable applications. This data, which was obtained from the report by Bernard Wolnak and Associates (8) is summarized in Table II. It shows that currently the projection is that the combined market for the major bulk enzymes, the amylases and proteases, will increase by less than 50% over the present decade, although within this broad market there will be some significant shifts and increases for individual enzymes.

TABLE II

Total U.S. Markets for Enzymes ($ million)

	1971	1975*	1980*
Amylases	8.31	12.50	14.20
Proteases	18.34	20.77	24.51
Glucose Isomerase	1.00	3.00	6.00
Cellulase	0.10	0.15	0.20
Glucose Oxidase	0.35	0.60	0.90
Other (Pectinase, Invertase, etc.)	1.66	1.85	2.10
Medical, Diagnostic, Research, etc.	5.50	7.30	9.80
Total	35.26	45.42	57.71

* Projected markets

Data from the report by Bernard Wolnak and Associates (8).

In contrast to the bulk enzymes the market for highly purified enzymes for use in the pharmaceutical industry, research, therapeutic application, diagnostics and clinical analysis is expanding rapidly and the prediction is that this market should double over the present decade. Also, the market for the individual enzymes glucose isomerase, cellulase and glucose oxidase will rise dramatically during this same period. Except for the sharply increased utilization of these individual enzymes and the promise presented by some of the more recent and novel applications it is hard, in

the light of the above assessments, to become highly excited about
the future growth prospects of the industry over the next few years.
However, recent concurrent advances in basic enzymology foretell
the possibility of a much more exciting and vital future for enzyme
technology and its applications, which is perhaps 5 to 10 years
away. But the question must be asked as to which industries will
benefit the most from these advances.

FUTURE TRENDS

Foremost among the recent important advances in basic enzymology is the practicality of immobilizing enzymes to soluble and
insoluble polymers and supports, but other advances in the areas of
enzyme purification by one-step processes, cofactor immobilization
and regeneration, genetic methods of increasing enzyme yields,
synzyme production and many others, are occurring simultaneously
and are threatening to revolutionize the production and use of enzymes during the next decade. But again the question must be
raised as to whether the present bulk enzyme industry will be the
primary beneficiary of these advances or whether the industry will
be hurt by them.

It is perhaps fortunate that these revolutionary advances are
occurring at a time in our history when we are being forced to recognize that we cannot continue much longer to run our energy production and organic chemicals industries on feedstocks of non-
renewable resources, especially oil. As we are now seeing, apart
from the associated political problems, these feedstocks are going
to cost significantly more as we attempt, because of declining re-
serves, to obtain these resources by importation and from less
accessible domestic origins, such as shale, coal and tar sands.
Consequently, it is almost certain that before the end of this
century supply restrictions and political and economic pressures
will force the conversion, wherever possible, from non-renewable to
renewable resources as the source of feedstocks for the chemical
industry. In essence this means that agricultural products and waste
products will become extremely important and will be required on a
massive scale. But how will we convert these complex plant products
into useful feedstocks with high efficiency? The ultimate answer
will surely be through the use of enzymes, the highly efficient and
selective catalysts which have evolved over many millions of years
specifically to deal with these materials. With some of the dark
predictions for our future now being made we should be thankful that
advances in enzyme engineering and technology are being made now,
hopefully in time for us to establish industries based on our renewable resources, the biologically-produced materials.

The recent dramatic advances in enzymology which have just
been mentioned and the concurrent surge of interest in enzyme technology both at the academic and industrial level, have been the
subject of a number of symposia and conferences over the past

several years. In particular, two five-day International Conferences on Enzyme Engineering have been held at Henniker, N.H. under the sponsorship of the Engineering Foundation Conferences. These conferences were held in the summers of 1971 and 1973 and readers are referred to the publications resulting from them for additional details of the topics covered (9,10). At the most recent conference over 50 scientific papers were presented on topics covering new and novel sources of enzymes, new purification techniques, immobilization techniques and supports, immobilized multi-step enzyme systems, physical methods for examining immobilized enzymes, immobilization and regeneration of coenzymes, enzyme reactor design, industrial applications of immobilized enzymes, commercial aspects of enzyme use and new applications of enzymes.

Some general impressions were reinforced by this conference. It was apparent that most industrial participants recognized the commercial importance of these new advances and were primarily concerned with economics, identifying immediate potential applications and also were considering the advantages of converting certain processes presently based on soluble enzymes over to the use of immobilized enzymes. But, in the majority of cases, they were primarily interested in the hydrolase enzymes since these show the greatest immediate potential. The academicians on the other hand were looking more to the future and considering the problems which will occur as greater and more sophisticated applications of other classes of enzymes such as the dehydrogenases, the hydroxylases and the synthetases are attempted. Foremost among these were the problems of coenzyme use and regeneration, the potentials of immobilized multi-step enzyme systems and the exciting possibilities of building useful metabolic pathways, fractions of complete metabolic pathways, or even totally new metabolic pathways, outside the cell on columns or in some other configuration. The problems of simple and economic purification of enzymes from cellular extracts were also intensively discussed. This latter point is very important since it is clear that the sophisticated processes now being considered for the future will require the more versatile chemistry of which only certain intracellular enzymes are capable.

Trends in Enzyme Production

Within the area of novel enzyme sources and enzyme production there is considerable interest in the methods now being developed by molecular biologists to bring about increased amounts of enzymes in cells. Among these techniques is the production of constitutive mutants which are cells having mutations in their regulatory genes. With inactive regulatory genes these cells produce the respective enzymes independent of inducible or repressible conditions. A potentially more valuable mechanism for increasing the amount of a specific enzyme in a cell is by the use of "superproducing" mutants which contain multiple gene copies. This latter technique can provide cells having 20% or more of their total protein as one enzyme. Such methods, once they can be applied to industrial processes should certainly ease problems of enzyme production and may also make enzyme purification and recovery simpler. However, it should be pointed out that productivity will not necessarily be increased by the same factor as the increase in enzyme content since these mutants usually grow more slowly than the wild type organism and consequently there will be problems with their culture. The other advantages will no doubt lead to their ultimate utilization in industrial processes.

Sources of enzymes other than the standard strains of microorganisms are now being carefully considered. It has been pointed out (11) that enzymes from thermophilic microorganisms generally are more heat-stable and have higher optimal temperatures than those from normal mesophilic organisms. Such enzymes have a number of features which make them potentially valuable in industrial applications. As summarized in Table III, enzymes from thermophilic organisms generally have a greater heat stability and higher optimal temperatures than the equivalent enzymes from mesophiles. These features have advantages in certain situations where it is essential to maintain high reactor temperatures because of low substrate solubility or to reduce viscosity. Also, a fact which is probably most important in food processing, higher reactor temperatures could reduce the degree of microbial contamination of the reactor. The observed decrease in sensitivity to denaturing agents such as solvents and detergents (12) might also prove to be of significant value in certain situations, while lower production costs because of some fermentation economies, improved yields on purification and greater stability on storage, all make it obvious that enzymes from thermophiles have great future potential in industrial processing.

There are, however, ways of obtaining heat-stable enzymes from the more common microorganisms. It was reported in 1970 by Isono (13) that *Bacillus stearothermophilus*, when grown at 55°, produces an α-amylase having greater heat stability than the α-amylase produced by the same organism grown at 37°. The α-amylase produced at 55° had both a greater heat stability (at 80° in the presence of $CaCl_2$), and a lower Km for starch (at both 65° and 37°) than the enzyme produced at 37°, although their electrophoretic mobilities and optimal temperatures appeared to be the same. However, preliminary data indicated differences in amino acid composition. Hopefully these fertile areas will soon be exploited by the industry.

The culture of human cell lines for the production of human enzymes and hormones of potential therapeutic value is an area which is now receiving increasing attention. Foremost among the products are urokinase for potential treatment of thromboembolic diseases and human insulin and growth hormone. The techniques of tissue

TABLE III

Potential Advantages in Industrial Processing of

Enzymes from Thermophilic Organisms

1. Greater heat stability

2. Longer stability on storage

3. Higher optimal temperatures

4. Better yields during purification

5. Higher enzyme reactor temperatures

6. Decreased enzyme production costs

7. Lowered sensitivity to denaturing agents

culture have progressed significantly over the past few years such
that it is no longer particularly difficult to culture many animal
cell lines. Substantial yields of enzymes and hormones can now be
achieved in relatively small and unsophisticated set-ups (14,15).

More into the future, synthetic polymers having enzyme-like
activity, the so called synzymes, might have an expanding role
but it is not yet possible to determine whether they, or the chemically synthesized enzymes, will have a viable industrial future.
Major advances will be needed before their future industrial importance can be assessed.

Trends in Enzyme Purification

There is now considerable interest in continuous enzyme purification processes, especially for intracellular microbial enzymes,
which could be incorporated into continuous enzyme production systems (16). London's University College has an operating system in
which microbial cells are produced by continuous culture, harvested,
broken and specific enzymes recovered and partially purified, mainly
by standard precipitation techniques, on a continuous basis (17).
Hopefully the enzyme industry will examine this lead and find it
economic and practical to produce purified enzymes on a large scale
by these methods.

It is highly probable that continuous purification methods
will be aided or even partially superseded by one-step procedures
such as affinity chromatography and immunoadsorption. These processes, which are capable of purification factors in the range of
several thousand in one step, might well significantly reduce the
cost of high purity intracellular enzymes if they were applied to

any appreciable extent in industry. Strangely enough industry does
not yet appear to be using these techniques to any degree, if at all,
despite their successful implementation at the bench level. Certainly, continuous affinity methods based on the use of toroidal
drums (17) should be of exceptional value to the producers of high
purity enzymes once the difficulties with this system have been
overcome. These appear to be primarily questions of obtaining
specific affinity ligands of suitable dissociation constants, developing suitable elution systems and overcoming the problems of un-
even flow rates and channeling caused by non-uniform packing of
the columns. However, even before these problems have been solved
completely, it is highly likely that affinity methods based on the
"tea-bag" approach will have industrial value, especially for the
semi-continuous recovery and purification of enzymes from crude
cellular extracts.

Trends in Enzyme Immobilization

        The recent development of the various techniques for immobi-
lizing enzymes, whether these be by adsorption, covalent bonding,
gel entrapment or microencapsulation, is perhaps the greatest single
impetus to the newly emerging era of enzyme technology and application. The benefits of enzyme immobilization, which include the
ability to recover and reuse enzymes, the use of enzyme columns or
membranes or similar configurations in continuous processes, the
frequent improvement in enzyme stability, the ease of removing enzymes from the final product by filtration or precipitation, and
potential decreases in process operating and capital costs, all
point to the future massive application of immobilized enzymes in
industrial processes.

        The extensive development of enzyme immobilization techniques
which is now occurring appears to be largely devoted to the design
of carriers which are relatively cheap, inert, have good mechanical
and flow properties for use in continuous systems and are capable
of being derivatized extensively. An additional feature which is
receiving attention is the potential for chemical modification of
the carrier surface. This allows the microenvironment of the bound
enzyme to be designed in such a way as to enhance or change the
enzymes kinetic properties. Activity-pH profiles and even the apparent kinetic constants of the enzymes can be successfully altered
by changes in the charged groups on the surface of the carrier (18).
Many of the more promising newer carriers have been designed with
these features in mind as well as the more obvious aspects of high
loading capacity and high surface area.

        Apart from the covalent bonding of enzymes to solid supports
(19) such as glass, cellulosic derivatives, modified nylon membranes
and imido-ester polymers, considerable interest is now being elicited
in the technique described by Dinelli (20) of wet spinning enzymes and
inert polymers together to obtain fiber entrapped enzymes.

These highly active enzyme fibers may then be used as monofilaments or woven in the form of cloth depending on the particular application required. It is reported (20) that many enzymes have been entrapped in this way and quite spectacular activities and stabilities have been achieved. This one advance may have exceptionally large applications in the food industry.

Another recent advance of considerable consequence to the food and fermentation industries is the immobilization of enzymes to magnetic powders (21). These have the advantage that they can be recovered from reaction vessels and product streams by simple magnetic separation techniques. The enzymatic processes which would benefit the most from this advance are those dealing with highly viscous or particle-containing fluids. Such processes are not readily applicable to column operation because of the problems of low flow rates and column plugging. However, they may be readily operated in continuous stirred tank reactors using enzymes immobilized on magnetic supports for continuous or batch recovery and reuse of the enzyme. The concept of using magnet supports may also have potential application for the recovery of enzymes from viscous or particle-containing extracts by affinity chromatography.

The potentials of immobilizing several different enzymes on a single surface are now being recognized. It has been shown (18) that multi-step enzyme systems immobilized on the same supporting surface can operate with very high efficiencies because intermediate products of the sequence diffuse only slowly from the Nernst layer and are consequently available to the next enzyme in the sequence at much higher concentrations than would be anticipated from their concentration in the bulk solution. However, the problems of different pH profiles for enzymes in such a sequence may limit the immediate potential of multi-step enzyme systems immobilized to single surfaces.

Not withstanding this, the future for such multi-enzyme sequences is very high and it might be predicted that they will eventually be used industrially to perform with high efficiency extensive chemical conversions of the types now found in cellular metabolism. Even more exciting is the prospect of being able to build completely novel chemical conversion routes by the use of presently unrelated enzymes in sequence. Such processes may well become the basis of a new industry for the conversion of carbohydrates - our renewable resources - to feedstock materials for our chemical industry. While this can already be done by fermentation methods, it might be anticipated that the use of enzymes would be much more efficient and would generate little in the way of by-products.

Trends in Coenzyme Applications

It was pointed out earlier in this paper that the first generation enzymes, which presently represent the overwhelming majority

of products of the enzyme industry, have essentially no cofactor
requirements. Equally well, these same enzymes only carry out the
relatively simple reaction of hydrolysis. The great benefits of
enzyme use, - their high specificity, high efficiency, low temperature operation and low level of side reactions - would be of greatest
benefit in fine chemical syntheses involving more complex reactions
such as specific hydroxylations, dehydrogenations and phosphoryla-
tions. While certain enzymes are superbly capable of performing
these reactions such enzymes invariably have an obligatory requirement for the low molecular weight pyridine or adenine nucleotide
coenzymes. In order to make use of these enzymes, as we will most
surely want to do in the more distant future, it will first be
necessary to find ways of retaining their respective coenzymes in
the reaction medium so that they are not lost in the product stream.
Also, we must discover ways of regenerating the coenzymes following
their involvement in the reaction.

     One possible solution to the problem of coenzyme retention
is to increase the molecular weight of the coenzyme by covalently
binding the native coenzyme to a soluble high molecular weight polymer and then using these coenzymes in reactors which rely on specific
cut-off membranes for separation of the products from the reaction
mixture. Of course, such a solution to the problem depends upon
the ability to covalently bind coenzymes to polymers in such a way
as to retain the reactivity of the coenzyme.

     Successful immobilizations of various coenzymes, including
NADH and coenzyme A have now been reported (22-24) and in some
cases have been shown to be catalytically active with native enzymes.
These advances hold the promise that the potentially more valuable
coenzyme-requiring enzyme systems may soon find significant industrial applications. They have not done so up till now primarily because coenzymes cost almost as much as purified enzymes and they
could not be recovered and regenerated effectively. Both of these
problems are now receiving attention and are being solved.

     The successful immobilization of various coenzymes is also
of great potential value in the recovery and purification of coenzyme-
requiring enzymes by affinity chromatography. Immobilized coenzyme
A has already been used to recover CoA-binding proteins which were
in turn immobilized to solid supports and used to recover CoA from
cellular extracts (24).

Trends in Industrial Applications

     It was stated previously that up until now the major indust-
rial applications of enzymes have involved the use of simple hydrolases - the first generation enzymes. One of the foremost applications is the use of α-amylase and amyloglucosidase in the production
of glucose syrups from corn starch. It is estimated that 1.3 billion

pounds of glucose was produced in the U.S. in 1971 and that the value of the enzymes used in its manufacture and that of corn syrup was well over $5 million (8). Alpha amylase is also used extensively in the paper and textiles industries where it is used principally to desize textiles and to solubilize starch for use in corrugated paper production and paper coatings. Such markets were over $4 million in 1971 (8).

The other major class of first generation enzymes is the proteases whose major applications are in cheese manufacture, as digestive aid additives in animal feeds, as additives in laundry detergents, as chill proofers in beer, as meat tenderizers and as gluten modifiers in baking.

With the dawning of a new era in enzyme technology, sparked primarily by the advent of immobilized enzymes, we are now entering a period of second generation enzyme application. In this period we are witnessing the development and introduction of processes based on immobilized hydrolases and other non-cofactor requiring enzymes. In many instances these are processes which were previously operated using soluble enzymes, but the conversion to immobilized enzymes has resulted in significant economies and other benefits. A typical case is that of the Tanabe aminoacylase process for the preparation of pure L-amino acids from racemic mixtures (25). In this process, which was the first industrial scale process to use immobilized enzymes, chemically synthesized racemic mixtures of individual α-amino acids, such as alanine or phenylalanine, are acetylated to form the acetyl-DL-amino acid. This mixture is then passed through a column of aminoacylase immobilized by adsorption to DEAE-Sephadex. The stereospecificity of the enzyme ensures that only the acetyl-L-amino acid is cleaved to yield the free L-amino acid. Evaporation and crystalization then allows the recovery of crystalline L-amino acid while the more soluble acetyl-D-amino acid is racemized and returned to the start of the process. Tonnage quantities of pure L-amino acids have been produced in this way. Previously the process was based on the batch treatment of acetyl-DL-amino acids with soluble aminoacylase, but the conversion to the immobilized enzyme has allowed the process to go continuous. This, together with economics resulting from the decreased need for enzyme and the greater overall efficiency of the process have lowered total costs by approximately 40%.

Another soluble enzyme process which may soon be converted to use immobilized enzymes is the production of glucose syrups from corn starch. Considerable development is presently underway on this immobilized enzyme process (26). Although the economies of conversion from the soluble enzyme process to the immobilized enzyme process have been questioned, the immobilized enzyme process may be of considerable significance as the first stage of a system to produce invert sugar from corn starch. This process requires the initial production of glucose from starch and then the partial conversion of glucose to fructose, using the enzyme glucose isomerase. In this latter stage the benefits of using immobilized glucose isomerase are clear, primarily because of the high cost of the enzyme and simpler continuous operation of the process. This process promises to have significant economic and perhaps political impacts because it provides the potential of replacing large amount of imported sucrose with a cheaper domestic product.

Many other processes important to the food industry may soon be converted to the use of immobilized enzymes. Included among these are the chill proofing of beer using immobilized papain, the clarification of fruit juices and wines using immobilized pectinase, the production of invert sugar from sucrose using immobilized invertase and the production of cheese using immobilized rennin. In many of these cases there are other advantages to using immobilized enzymes, other than the potential reduction in enzyme costs. In food processing, the use of immobilized enzymes is especially important since it assures that the enzyme does not remain in the final product. This allows greater control of the process and may even overcome problems with certain government regulations.

One process in which this may be of fundamental importance is in the treatment of milk with lactase to hydrolyze the milk sugar lactose. In most countries around the world, apart from those countries inhabited primarily by populations of northern European ancestry, the vast majority of the adult population is lactose intolerant. This means that they cannot digest lactose and consequently, any non-fermented milk products in their diet can cause considerable intestinal problems and discomfort. One answer to this problem, especially where feeding of milk products would raise the quality of the diet, is to pretreat the milk with the enzyme lactase to convert lactose to the readily absorbed sugars galactose and glucose. The simplest and probably the cheapest way to treat the milk would be by using columns packed with immobilized lactase. Certain difficulties with this process must still be solved, especially the problem of microbial growth on the columns, but when they are the large scale application of this process will have significant social and economic impacts in many countries, and even in the U.S. where approximately 60% of the adult black population is lactose intolerant. Furthermore, this same process could be used for the treatment of whey, the by-product of cheesemaking. Reduction in the concentration of lactose in whey would allow more whey to be used in other food products, such as icecream and protein supplements.

CONCLUSION

The present trends in enzyme technology and application are reasonably clear. Certain processes using bulk hydrolases will probably continue along very much the same way as they have always

been used, primarily because there is little economic advantage to
converting to immobilized enzymes or, even more important, because
of the practical impossibility of using recoverable immobilized
enzymes. Among these processes are meat tenderization and gluten
treatment of flour with proteases. Other enzyme applications may
be slightly modified by the new enzyme technology, with one example
being the possibility of microencapsulating detergent enzymes.

Within the realm of the first generation enzymes there is a
growing trend towards the search for more novel hydrolases which have
highly specific applications. Here we can think of melibiase, tannase, keratinase, etc.

However, the major trend we can expect over the next few
years is towards the increased application of the second generation
enzymes, the immobilized non-cofactor requiring enzymes. Once the
first few major industrial applications of these enzymes occur and
the economic advantages become well recognized we can expect a
rising crescendo of such applications to appear.

In the more distant future, perhaps 10-20 years away, we
can see the development and application of completely new processes
based on a third generation of enzymes, those which require coenzymes and coenzyme regeneration systems. These enzymatic processes,
which will be based on research currently being performed primarily
in academic institutions on coenzyme immobilization and regeneration,
and immobilized multi-step enzyme systems, may well revolutionize
synthetic chemistry and the industrial production of pharmaceuticals
and bulk chemicals. Some of these processes, such as chenodeoxychol-
ate synthesis from cholic acid and the bulk production of potentially
important dicarboxylic acids from phenol using immobilized enzymes
and coenzymes are already being examined by our own research group
at the University of Pennsylvania. Other groups around the world
are also actively pursuing similar goals, but unfortunately it may
be assumed that a significant time gap will occur between the demonstration of the feasability of such processes and their implementation at the industrial level. This gap may, however, be shortened
in response to the pressures produced by increased costs of non-
renewable resources.

Thus, the extensive use of immobilized enzymes in industrial
processes and analytical devices, which is another area with spectacular possibilities, together with the application of other recent
advances, can confidently be expected to change the face of various
industries, including the food, pharmaceuticals, chemicals and the
enzyme industry itself.

REFERENCES

1. German Patent No. 283923.
2. W. M. Fogarty and O. P. Ward, *Process Biochemistry 7*, 13 (1972).
3. Tanabe Seiyako Co., Ltd., Osaka, Japan.
4. "Achievement in New Beet Sugar Producing Method by Using Newly Discovered Enzyme". Fermentation Research Institute, Agency of Industrial Science and Technology, Tokyo, Japan, 1969.
5. Kyowa Hakko Kogyo Co., Ltd., Tokyo, Japan. *News Fair No. 3* (1971).
6. E.T. Reese, M. Mandels and A.H. Weiss. In *Advances in Biochemical Engineering, Vol.2* (Eds. T.K. Ghose, A. Fiechter and N. Blakebrough). Springer-Verlag, Berlin, 1972, p.181.
7. A. J. Wieg, *Process Biochemistry 5,* 33 (1970).
8. "Present and Future Technological and Commercial Status of Enzymes." Bernard Wolnak and Associates, Chicago, Illinois, 1972.
9. L. B. Wingard, Jr. (Ed.) "Enzyme Engineering", John Wiley, New York, 1972.
10. E.K. Pye and L.B. Wingard, Jr. (Eds.) "Enzyme Engineering Vol. 2" Plenum Press, New York, (in the press).
11. A. R. Doig. In "Enzyme Engineering Vol. 2" (Eds. E.K. Pye and L.B. Wingard, Jr.) Plenum Press, New York (in the press).
12. J. Farell and L.L. Campbell, *Adv. Microbial Physiol. 3,* 83 (1969).
13. K. Isono. *Biochem. Biophys. Res. Commun. 41,*852 (1970).
14. M. Posner. In "Enzyme Engineering Vol. 2" (Eds. E.K. Pye and L.B. Wingard, Jr) Plenum Press, New York (in the press).
15. L. K. Nyiri. In "Enzyme Engineering Vol. 2" (Eds. E.K. Pye and L.B. Wingard, Jr.) Plenum Press, New York, (in the press).
16. P. Gray, P. Dunnill and M.D. Lilly. In "Fermentation Technology Today", (Ed. G. Terui) Soc. Fermentation Technol. Japan, 1972, p. 347.
17. P. Dunnill and M.D. Lilly. In "Enzyme Engineering" (Ed. L.B. Wingard, Jr.) John Wiley, New York, 1972, p.97.
18. K. Mosbach, B. Mattiasson, S. Gestrelius and P.A. Srere. In "Enzyme Engineering Vol. 2" (Eds. E.K. Pye and L.B. Wingard, Jr.) Plenum Press, New York, (in the press).
19. O.R. Zaborsky. "Immobilized Enzymes", Chemical Rubber Company Press, Cleveland (1973).
20. D. Dinelli and F. Morisi. In "Enzyme Engineering Vol. 2" (Eds. E.K. Pye and L. B. Wingard, Jr.) Plenum Press, New York (in the press).
21 P.J. Robinson, P. Dunnill and M.D. Lilly. *Biotechnol. Bioeng. 15,* 603 (1973).

22. M.K. Weibel, C.W. Fuller, J.M. Stadel, A.F.E.P. Buckmann,
      T. Doyle and H. J. Bright. In "Enzyme Engineering Vol. 2"
      (Eds. E.K. Pye and L. B. Wingard, Jr.) Plenum Press, New
      York (in the press).
23. K. Mosbach, P.-O. Larsson, P. Brodelius, H. Guilford and M.
      Lindberg. In "Enzyme Engineering Vol. 2" (Eds. E.K. Pye
      and L.B. Wingard, Jr.) Plenum Press, New York, (in the
      press).
24. I. Chibata, T. Tosa and Y. Matuo. In "Enzyme Engineering Vol.2"
      (Eds. E.K. Pye and L.B. Wingard, Jr.) Plenum Press, New
      York (in the press).
25. I. Chibata, T. Tosa, T. Sato, T. Mori and Y. Matsuo. In
      "Fermentation Technology Today" (Ed.G. Teru) Society of
      Fermentation Technology, Japan, 1972, p. 383.
26. H.H. Weetall. *Food Prod. Develop.* 7, 46 (1973).

IMMOBILIZED ENZYMES IN MILK SYSTEMS

T. Richardson and N. F. Olson

Department of Food Science, University of Wisconsin

Madison, WI 53706

Milk systems offer ideal media for studying the use of immobilized enzymes in food processing and analyses. For example, milk and many milk products are fluids which facilitate their processing with immobilized enzymes. Furthermore, some processes in the dairy industry already require the use of enzymes, e.g. the clotting of milk by rennin, pepsin and the microbial milk-clotting enzymes. Thus, there are excellent opportunities for integrating immobilized enzymes into present dairy processing situations, possibly in a continuous mode of operation. In addition, modification of milk constituents, such as milkfat with immobilized lipases, offers further opportunities for processing.

Our laboratories have been engaged in research on immobilized enzymes for processing and analyses in five areas related to the dairy industry. These include: 1) immobilized catalase for the destruction of hydrogen peroxide (H_2O_2) used in the "cold sterilization" of milk, 2) immobilized peroxidases as antimicrobial agents, 3) immobilized proteases for coagulation of milk during cheese manufacture, 4) immobilized papain for studying the structure of the casein micelle, and 5) immobilized β-galactosidase for the hydrolysis of lactose in dairy products. In the following discussion, each of the above areas of research will be discussed separately.

IMMOBILIZED CATALASE

Catalase is used in the food industry to destroy unwanted hydrogen peroxide resulting from processing operations. In the dairy industry, up to 0.05% H_2O_2 has been used to "cold pasteurize"

19

milk being prepared for the manufacture of Cheddar and related
varieties of cheese and Swiss cheese. After "cold pasteurization"
is completed, excess H_2O_2 is destroyed with catalase. However, use
of H_2O_2 to treat milk has been rather limited due, in part, to the
cost of the enzyme and to the inconvenience of using catalase to
destroy residual H_2O_2 in milk before subsequent processing. Thus,
an immobilized catalase might prove useful as a convenient and
economical means of removing H_2O_2 from treated milk.

We have immobilized catalase a variety of ways including cross-
linking crystalline catalase (Ferrier et al., 1972), entrapment in
dialysis tubing, adsorption on DEAE-cellulose or cheesecloth and
subsequent cross-linking with glutaraldehyde, and coupling directly
to cheesecloth previously oxidized with periodate (Balcom et al.,
1971). The DEAE-cellulose-catalase and cross-linked catalase crystals exhibited good activity with the other preparations possessing
lesser activities. However, the immobilized catalase was not very
stable during continuous usage as shown in Figure 1. Catalase
cross-linked on cheesecloth was fairly stable for only the first 40
minutes of operation. Activity decreased rapidly during subsequent
use. The lack of stability has been reported also for soluble
catalase in the presence of higher levels of H_2O_2 (George, 1947;
Miller, 1958; Morgulis et al., 1926).

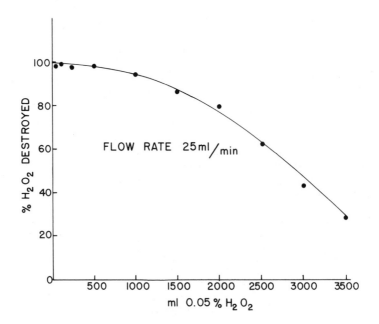

Figure 1. Inactivation of catalase, cross-linked on cheesecloth
with glutaraldehyde, during continuous treatment of H_2O_2.

The problem of instability may be overcome partially by use of different immobilization techniques or by use of the optimum reactor design. O'Neill (1972) recently reported that the rate of immobilized catalase inactivation may be reduced considerably by using a continuously stirred tank reactor (CSTR) rather than a bed reactor if hydrogen peroxide destruction follows zero order kinetics. He concluded the catalase inactivation in CSTR would be no slower than a packed bed reactor if the reaction were first order.

IMMOBILIZED PEROXIDASES

A number of studies have implicated peroxidases as antimicrobial agents. The neutrophil peroxidase, myeloperoxidase, exerted antibacterial effects against Escherichia coli and Lactobacillus acidophilus when combined with hydrogen peroxide and either thiocyanate ions (Klebanoff et al., 1966; Klebanoff and Luebke, 1965) or a halide such as iodide, bromide or chloride ions (Klebanoff, 1967a; Klebanoff, 1967b; Klebanoff, 1968). In addition, purified human myeloperoxidase in the presence of iodide and hydrogen peroxide exerted killing activity against cells of several species of Candida, Saccharomyces, Geotrichum and Rhodotorula and spores of Aspergillus fumigatus and A. niger (Lehrer, 1969). This system is thought to be a natural bactericidal component of leukocytes (Klebanoff, 1972; Simmons and Karnovsky, 1973). Furthermore, antibacterial systems based on peroxidases appear to be present naturally in milk and saliva (Hogg and Jago, 1970; Imamoto et al., 1972; Klebanoff and Luebke, 1965). Apparently, peroxidase in the presence of hydrogen peroxide and iodide exerts its antibacterial effect by iodination of bacteria (Klebanoff, 1967b).

A major problem in the use of immobilized enzymes for processing complex biological materials such as food is the proliferation of microorganisms in the enzyme reactor (Ferrier et al., 1972). Columns of some immobilized enzymes can be sanitized with dilute hydrogen peroxide (Ferrier et al., 1972). However, it would be desirable to have an antimicrobial system active in situ to prevent the multiplication of microorganisms during processing.

Consequently, we have initiated studies on the bactericidal properties of immobilized peroxidases which may be useful as antimicrobial agents in enzyme reactors or in the "cold sterilization" of fluids.

In our studies, a heme-proteinoid (HP) with peroxidatic activity (No. 83a [Dose and Zaki, 1971]) was prepared by the thermal polymerization of the appropriate mixture of amino acids and bovine hemin. The heme-proteinoid (HP), lactoperoxidase (LP), horseradish peroxidase (HRP) and bovine serum albumin (BSA) were immobilized on Sepharose 4B activated with cyanogen bromide (Cuatrecases et al., 1968).

Peroxidatic activity was determined from the oxidation of guaiacol and arbitrary guaiacol units were assigned to each preparation (Rosoff and Cruess, 1949). Bactericidal activities of the immobilized peroxidases were measured using Escherichia coli and Staphylococcus aureus (strain 100). Eighteen- to twenty-hour cultures of the microorganisms were diluted to give approximately 2×10^6 organisms per ml of reaction mixture in a total volume of 10 ml. A blank reaction mixture contained only organisms in 0.01 M phosphate buffer at pH 6.5. All treatment tubes included 8.6 x 10^{-4} M H_2O_2 (concentration determined by iodometric titration), and 0.001 M potassium iodide. Control tubes contained the peroxide-iodide mixture alone and also in the presence of 1.8 mg of immobilized BSA to test for nonspecific effects of protein and Sepharose. In the flasks containing peroxidases, 0.026 units of LP-Sepharose (0.2 mg), 2 units of HRP-Sepharose (1.8 mg) and 0.19 units of HP-Sepharose (1.7 mg) were used. The volumes in all tubes were adjusted to 10 ml with phosphate buffer, pH 6.5. The tubes were incubated at 25° for 30 min with occasional shaking. Appropriate dilutions of the various mixtures were then plated on trypticase soy agar (with 0.5% lactose for E. coli) and viable cells counted using the standard plate count procedure of the American Public Health Association (1967).

The data in Table I indicate that treatment with hydrogen peroxide alone and in the presence of immobilized BSA produced very low bactericidal activity. The slight kill in the case of S. aureus is rather surprising in view of the low levels of H_2O_2 and KI. The LP-Sepharose caused a significantly greater destruction of both bacterial species, whereas the HRP-Sepharose and HP-Sepharose exhibited progressively lower bactericidal effects. However, the amount of enzyme used in various treatments must be taken into consideration in evaluating bacterial kill. A column or other reactor containing more enzyme and under turbulent conditions would tend to maximize the bactericidal effects. Furthermore, the level of HRP-Sepharose peroxidatic activity was approximately 80-fold greater than the activity of LP-Sepharose yet the latter enzyme exhibited much higher bactericidal effectiveness. The level of peroxidatic activity of the HP-Sepharose was also low. This suggests that higher levels of HP-Sepharose might make it feasible as a bactericidal agent. Since HP-Sepharose can be prepared fairly simply, the economics of using it in the immobilized form may be favorable.

A combination of glucose and glucose oxidase has been used to generate the H_2O_2 in situ for peroxidatic activity (Klebanoff, 1967a). Thus a system of immobilized glucose oxidase and immobilized peroxidatic agents might prove useful for bactericidal purposes. Other halides such as the ubiquitous chloride ion are

Table I. Bactericidal Effectiveness of Peroxidases against
Escherichia coli and Staphylococcus aureus

	Percentage kill[a]		Peroxidatic activity (guaiacol units)	
	S. aureus	E. coli	units/test	units/mg
Blank	0	2.7 ± 1.0		
H_2O_2-KI	29.1 ± 6.4	0		
BSA-Sepharose	28.9 ± 7.0	0		
LP-Sepharose	85.0 ± 6.2	84.8 ± 4.2	0.026	0.14
HRP-Sepharose	40.1 ± 0.5	5.0 ± 5.0	2.0	1.14
HP-Sepharose	51.0 ± 7.0	16.0 ± 5.4	0.19	0.11

[a]Average of two experiments ± the deviation from the mean. Each
experiment had three replicates.

effective in this system (Agner, 1972), however, the relative
effectiveness of chloride with immobilized peroxidases is yet to
be determined.

The bactericidal properties of immobilized peroxidases might
be useful in preventing the proliferation of microorganisms in
reactors involving other immobilized enzymes. Alternatively, this
system might be developed to continuously "cold sterilize" fluids
such as foods and medical fluids.

Preliminary results indicate no effect of any immobilized
peroxidases on the germination of Bacillus megaterium spores (NRRL
B 1368).

IMMOBILIZED PEPSIN

Immobilized proteases might be valuable in continuous coagulation of milk for cheese manufacture. Since the immobilized
enzyme would not remain in the product, it may be possible to sub-
stitute a less expensive, less desirable, but more readily available enzyme which normally cannot be used, such as crude microbial
proteases, instead of commercially available milk-clotting enzymes.

The coagulation of milk by a column or other reactor containing immobilized proteases is made possible by the behavior of milk
at low temperatures. Milk coagulation can be divided into two

phases: enzymic action on k-casein, and the subsequent gelation or clotting. Rate of clotting decreases 15- to 20-fold by lowering the temperature 10° (Ernstrom, 1965) whereas enzymic activity of pepsin-glass has a temperature coefficient of about 1.5 (Line et al., 1971). By lowering the temperature of the enzyme bed it would be possible to retain sufficient enzyme activity and yet prevent coagulation of the skimmilk until after emergence from the reactor. Subsequent warming of the milk would cause rapid clotting. The curd could then be processed continuously.

Attempts have been made to immobilize rennin (Green and Crutchfield, 1969) and chymotrypsin (Dolgikh et al., 1971; Green and Crutchfield, 1969) for coagulation of milk but the immobilized enzyme activity was very low (Dolgikh et al., 1971) or soluble proteolytic activity continued to leach from the immobilized enzyme preparations (Green and Crutchfield, 1969) thus preventing a definitive study.

Our work on milk coagulation has involved the use of pepsin immobilized on porous glass beads. Pepsin covalently coupled to 40- to 60-mesh porous glass was prepared (Line et al., 1971) and generously supplied by Dr. H. H. Weetal of Corning Glass Works, Corning, New York. Two samples of pepsin-glass supplied contained 22 and 41 mg protein per g glass as estimated from the arginine content. The first sample was prepared using crystalline pepsin and the second using crude pepsin so that the second sample (41 mg protein per g glass) had less enzymic activity. Columns of pepsin-glass of the desired height were prepared by pouring the glass particles into jacketed columns which were 0.9 by 30 cm and equipped with porous Teflon support discs. The columns of enzymes were washed with 100 ml of water. The jacketed columns were adjusted between 5° and 20° which separated the primary and secondary phases of skimmilk clotting. Cold acidified skimmilk flowed in a thin film down the side of the column (about 20 cm) before entering the enzyme bed. A pressure head of up to 60 cm was imposed on the enzyme bed to attain desired flow rates. Relative enzyme activities of the column of pepsin-glass were estimated by determining time for effluent skimmilk to coagulate at 30° with a Sommer-Matsen rennet tester (Ernstrom, 1965). Effluent skimmilk was collected in a 10-ml graduated cylinder cooled in an ice-bath and duplicate 2-ml samples were added quickly to prewarmed 125-ml bottles with a pipette chilled in ice. Timing of coagulation started immediately after the skimmilk was blown from the pipette.

The absence of soluble activity leaching from the glass beads was determined in two ways. The columns of enzyme were washed with simulated milk ultrafiltrate, and the washings were shown not to have proteolytic activity against casein. In addition, whey from milk coagulated by immobilized pepsin was not proteolytic toward added casein.

Microbial Growth on Columns

Columns of pepsin-glass became contaminated with microorganisms during use. When skimmilk was passed through a column and the column was washed with water, a small amount of white material remained on the column and under quiescent conditions it supported microbial growth. To test for microbial growth, a used 5.5-cm column stood under distilled water at room temperature for 48 hr and then was washed with 500 ml of sterile water. Both the pepsin-glass particles and the water contained microorganisms. Shaking the same pepsin-glass with five 100-ml portions of sterile water removed substantial numbers of microorganisms from the pepsin-glass with each wash but the glass particles still retained microorganisms. Treatment of columns of pepsin-glass with 0.05 M H_2O_2 for 30 min at 15° effectively sterilized the columns without reducing the clotting activity. Thus, microbial hazards could be minimized by washing the columns with H_2O_2 solutions and by use of properly pasteurized milk.

Stability of Insoluble Pepsin During Storage

Line et al. (1971) found that, in the moist form, pepsin-glass retained 100% activity after 30 days at 6°. In this study, the insoluble pepsin retained essentially the same clotting activity when stored for several months at 4°. A further point to consider is the enzyme stability when stored wet or even under water for several hours each day as might be the case when an industrial column of enzyme is not in use. Therefore, about 100 g of pepsin-glass was washed with 350 ml of 0.05 M H_2O_2 for 30 min to sterilize it and then it was divided into three portions and each was stored in water at 5°, 15°, or 25°. Duplicate 0.9 by 5.0 cm columns of the stored samples of insoluble pepsin were tested for clotting activity at intervals. Changes in enzymic activity during 4 weeks of storage are shown in Figure 2. The sample stored at 25° may have lost activity due to microbial contamination which occurred in this case; however, a pepsin-glass sample stored under 0.1% H_2O_2 for 1 week at 25° lost activity at a slightly greater rate. After 24 days, water in the 25° sample had a pH of 7.5, which probably inactivated the enzyme. The pH of the other two samples was 6.1. The cause for the increase in activity of the 5° and 15° samples is unknown.

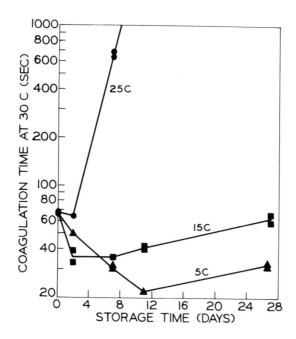

Figure 2.   Stability of pepsin-glass during storage under water
at 5°, 15°, or 25°.

Effect of pH on Coagulation Time

　　　Skimmilk acidified to various pH values between 5.3 and 6.7,
was passed through 0.9 by 5.5 cm columns of pepsin-glass at 15°.
After 20 min of flow at a rate of 6.0 ml/min, coagulation times of
samples of the effluent skimmilk were measured at 30°.   The effect
of varying pH of skimmilk on coagulation time of effluent skimmilk
is shown in Figure 3.   The break in the pH curve at about pH 6.2
was below the pH at which inactivation of pepsin occurs (Ernstrom,
1965).   A similar curve was obtained when soluble rennet was used
(Green and Crutchfield, 1969) even though rennet is stable up to
pH 7.5.   Therefore, the pH profile obtained is probably a characteristic of milk coagulation rather than the effect of pH on
pepsin-glass.   In any event, the effective operating pH for skimmilk coagulation was less than pH 6.1.

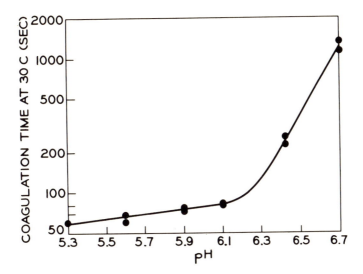

Figure 3. Hydrogen ion profile of skimmilk coagulation with pepsin-glass.

Immobilized pepsin might be integrated into a continuous system of manufacture of Mozzarella cheese (Larson et al., 1970; Quarne et al., 1968). To simulate conditions of this method, skimmilk was adjusted to pH 5.6 before treatment with pepsin-glass. However, maintaining flow rates for 10 to 12 hr proved difficult. This problem was partly overcome by raising the pH to 5.9 although coagulation times of the effluent skimmilk increased somewhat. As expected, the curd resulting from skimmilk at pH 5.9 was tougher and less extensible than that produced from skimmilk at pH 5.6

Operational Lifetime of Pepsin-Glass

An immobilized enzyme must remain active during a long period of use to be economically feasible. The operational lifetime of pepsin-glass was influenced by four factors: temperature, the sample of enzyme, inactivation during use, and accumulation of a white solid in the interstices of the column and the glass particles.

Plugging of the Columns. A white particulate material, similar in appearance to coagulated milk, accumulated in the interstices of the column during the passage of skimmilk through a column of pepsin-glass. This material reduced the flow since a gradual increase in head pressure (maximum head was about 60 cm) was required to maintain a constant flow. The tendency for columns of pepsin-glass to plug was influenced by both the sample of pepsin-glass and the pH. A flow of 6.0 ml/min could be easily maintained for more than 12 hr with skimmilk at pH 5.9 and pepsin-glass prepared with purified pepsin. If skimmilk at pH 5.6 was used, flow dropped below 6.0 ml/min in 8 hr or less, even with a pressure head of 60 cm. Flow rates decreased more rapidly when pepsin-glass prepared with crude pepsin was used. Although skimmilk at pH 5.9 was used in this case, flow rates of 6.0 ml/min proved difficult to maintain for more than 8 to 10 hr.

Higher flow rates reduced the tendency for columns to plug, but the clotting activity was reduced disproportionately at the highest flow rate (Table II). Preliminary experiments with pepsin-glass columns 50 cm long and with a flow of 50 ml/min indicated that such a column may be operated in excess of 12 hr without appreciable increase in head pressure. Accumulation on the pepsin-glass of the white material appeared to be reduced markedly.

Inactivation of Pepsin-Glass by Skimmilk. Continuous passage of skimmilk through columns of pepsin-glass resulted in gradual reduction in enzymic activity as measured by the rate of coagulation of the effluent skimmilk. As shown in Figure 4, the rate of loss of enzymic activity depended on the sample of pepsin-glass used. Pepsin-glass prepared with purified pepsin (Sample 1, upper curve) was inactivated more rapidly than the pepsin-glass prepared from crude pepsin (Sample 2, lower curve). The inactivation rate was the same for Sample 1 whether skimmilk was at pH 5.6 or 5.9. Sample 2 was tested only at pH 5.9. Either high or low flow rates resulted in about the same rates of inactivation of Sample 2 of pepsin-glass.

The pepsin-glass columns were completely stable to simulated milk ultrafiltrate at pH 5.6 or 5.9. However, passage of whey through the columns tended to inactivate them.

Table II. Effect of Increased Flow Rates on
Coagulation Times of Effluent Skimmilk from
a 5.0-cm Pepsin-Glass Column

Flow	Coagulation at 30°
ml/min	sec
3	44
5	60
7	93
9	169

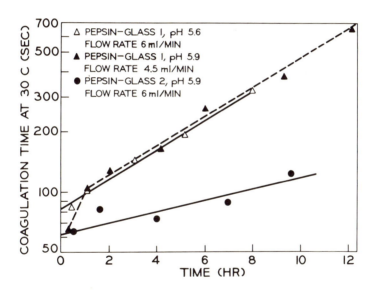

Figure 4. Loss of enzymic activity during continuous flow of
skimmilk through a pepsin-glass column at 15°.

The white material which accumulated in the interstices of the column during skimmilk flow could be removed easily by backwashing the column with water. However, after washing the column, the pepsin-glass particles remained coated with a material which resembled coagulated milk. To investigate the nature of this material, columns of pepsin-glass used for various times up to 8 hr were backwashed and the pepsin-glass particles were analyzed for nitrogen (Kjeldahl). Exposure of the pepsin-glass to skimmilk resulted in rapid increase in bound nitrogen to about four times the original amount. The rate of increase did not correspond to the rate of loss in clotting activity (Figure 5). Analysis of the material eluted from beads suggested that it contained a high proportion of peptides. The fraction most tenaciously bound to the glass contained high levels of sialic acid; the ratio of sialic acid to nitrogen was great enough to indicate that a portion of this fraction may be the glycomacropeptide released from k-casein.

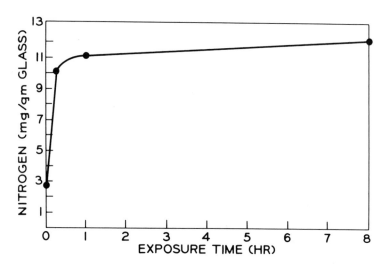

Figure 5. Accumulation of nitrogenous materials on pepsin-glass during continuous flow of skimmilk through the enzyme column.

Reactivation of Pepsin-Glass

Columns of pepsin-glass were reactivated after use by washing the column with 0.005 M HCl. This treatment effectively removed the white particulate material from the enzyme bed. The column stood overnight at about 25° and was washed intermittently with water which reactivated enzyme columns. Elution of the bound material with 4 M urea inactivated the enzyme. Columns of pepsin-glass which were either washed with dilute HCl before use or regenerated with dilute HCl after use were more active than the original column but, on subsequent use, they lost their enzymic activity at a faster rate (Figure 6). Thus, regeneration with dilute HCl seems impractical. Presumably further digestion of white material in the enzyme bed allowed its gradual removal when a column stood in water at 25°. Higher temperatures may allow faster reactivation of the enzyme. Optimum conditions for regeneration by digestion are under further investigation.

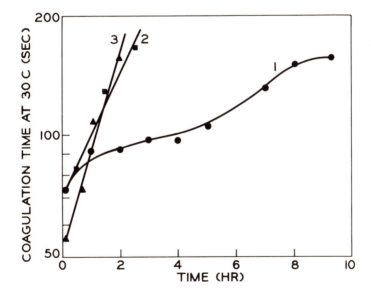

Figure 6. Loss of activity during repetitive use of a 50-cm pepsin-glass column. Column was used once (Curve 1) and then treated with 0.005 M HCl before each of two successive trials (Curves 2 and 3).

General Discussion

In practice, the use of an immobilized enzyme for continuous coagulation of skimmilk depends on several critical points, including separation of the enzymic and clotting stages so that clotting does not occur in the enzyme bed, high enzymic activity, sufficiently long retention of enzymic activity under operating conditions (possibly combined with the ability to be regenerated), freedom from microbial hazards, and the production of a normal product.

In this study, separation of the enzymic and clotting stages was accomplished by proper temperature control. The pepsin-glass had relatively high activity; sufficient enzyme was bound to cover the glass surface (40 m^2/g glass) with an enzyme layer 1 molecule thick (Line et al., 1971). It required about 0.2 g of pepsin-glass per milliliter milk per minute at 15° to produce effluent skimmilk which coagulated in 60 to 70 sec at 30°. Since pepsin-glass could be sterilized with 0.05 M hydrogen peroxide, use of pepsin-glass is not expected to result in increased microbial hazards. This area is being investigated further. High flow rates with turbulent flow should reduce the problem of column plugging.

The reduction in enzymic activity of pepsin-glass during use is a problem whose cause and remedy remain obscure. Possible approaches to the problem are to use a different enzyme which is inactivated much more slowly or not at all, or, to find a method for reactivation of the pepsin-glass which does not result in a more rapid inactivation of the enzyme upon reuse. It is possible that another immobilized enzyme rather than pepsin-glass would be preferable for continuous coagulation of skimmilk. Use of pepsin has the disadvantage that it is inactivated above pH 6.5 (Ernstrom, 1965) so that its use for continuous coagulation of milk at the normal pH of milk is not feasible. Even the pH of acidified skim-milk is considerably above the pH optimum of pepsin. Although porous glass has a high surface area (40 m^2/g) the average pore size is only 55 nm which is smaller than the diameter of many milk micelles (McMeekin, 1965). Thus, there is much potential enzymic activity which is probably not being utilized unless milk coagulation is caused by enzymic action on k-casein which is not on the casein micelle (Parry and Carrol, 1969). Porous glass also readily accumulated (colloidal) material from skimmilk which was deleterious to its use in enzyme columns although use of high flow rates may circumvent this problem.

The fact that immobilized pepsin can cause milk to coagulate implies that a portion of the k-casein is either in solution or on the surface of the casein micelle.

Immobilized milk-clotting enzymes could be incorporated into a continuous cheese manufacturing system as shown in Figure 7. In this procedure, milk flows through a smooth-bore tubular system. Food grade acids are injected into the milk stream at 5° to adjust the pH of milk to the desired value (Quarne et al., 1968). The milk passes through the immobilized enzyme reactor and is warmed to 35°. Coagulation of milk and expulsion of whey from curd occurs under turbulent conditions created by controlled flow rates through the tube (Larson et al., 1970). Curd and whey are separated continuously as they emerge from the tube. The curd is then formed into its final shape.

Use of immobilized milk-clotting enzymes could give greater flexibility and control over cheese ripening and merchandising. The milk-clotting enzymes, a major factor in initial hydrolysis of protein in cheese, would not contaminate the cheese. Therefore, addition of proteases could be varied to attain desired rates of proteolysis in cheese. The rubbery, elastic characteristics of cheese could be maintained longer during ripening by adding less

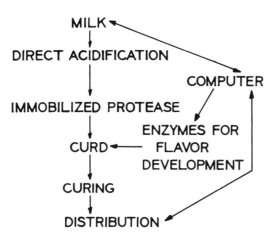

Figure 7. Flow diagram of a continuous cheese manufacturing process using immobilized milk-clotting enzymes.

enzyme. This can be used to advantage in manufacturing and merchandising a cheese variety like Mozzarella which cannot be stored
for extended periods. Excess cheese could be manufactured when
milk supplies are plentiful; the cheese held for extended periods
of time and merchandised when desired. In contrast to the above
situation, larger quantities of proteases and other enzymes can be
added to accelerate cheese ripening.

The system could be controlled and monitored by computers as
shown in Figure 7. Market reports are fed back to adjust the
amount of cheese manufactured and the amounts and types of ripening enzymes used. Price projections of milk and cheese could also
be used to control the addition of enzymes.

Use of immobilized enzymes may be advantageous also in separating the milk-clotting and cheese-ripening functions of proteases.
Choice of traditional milk-clotting enzymes is a compromise between
their milk-clotting and proteolytic activities. Immobilized proteases could be selected for their milk-clotting activity and for
effects on cheese yield and characteristics of cheese curd. Soluble proteases would be used for their effects on cheese ripening.

IMMOBILIZED PAPAIN

Our laboratories have used immobilized papain to study the
structure of bovine casein micelles (Ashoor et al., 1971). The
major casein fractions, α_s, β and k, are arranged in phosphate-
citrate complexes to form micelles which are 300-3,000 Å in diameter. A number of models have been proposed to define the arrange-
ment of the casein fractions in the micelle (Garnier and Ribadeau-
Dumas, 1970; Kirchmeier, 1970; Parry and Carrol, 1969; Payens,
1966; Ribadeau-Dumas and Garnier, 1970; Rose, 1969). For example,
some researchers suggest that k-casein is located on the surface
of the micelle as a stabilizing coat, whereas others believe the
casein fractions are distributed uniformly throughout the micelle.

Since the micelles have been shown to be very porous, accommodating enzyme molecules of 35,000 daltons in the interior
(Ribadeau-Dumas and Garner, 1970), the number of reagents suitable for studying surface and internal micellar structures is
limited. To circumvent this problem, a superpolymer of papain was
prepared by polymerization of papain with glutaraldehyde. This
protease polymer would not penetrate the micelle and could be used
to selectively digest the exterior portions of the micelles.

Micellar and soluble casein samples were treated with papain which had been cross-linked with glutaraldehyde to form a papain superpolymer. Unhydrolyzed portions of casein samples were fractionated on polyacrylamide gels and the stained gels were scanned with a densitometer. In both casein samples, the amount of unhydrolyzed k-, β-, and α_s-caseins decreased gradually as the reaction proceeded. The percentage of hydrolysis after 60 min of reaction was 69% for micellar casein and 78% for soluble casein. However, none of the three casein fractions was hydrolyzed completely. Unhydrolyzed casein in both samples had approximately the same composition throughout the entire reaction. These results suggest that k-casein does not have a specific location in the casein micelle and that the three major casein fractions are distributed uniformly throughout the micelle.

IMMOBILIZED β-GALACTOSIDASE

Part of our work with milk systems has involved immobilization of β-galactosidase from Escherichia coli K-12 by a relatively unique method (Hustad et al., 1973a; Hustad et al., 1973b). The enzyme was coupled to a polyisocyanate polymer which had been molded onto a carrier. In our case, the carriers were magnetic stirring bars. The coupling reaction is illustrated in Figure 8. A commercial preparation of polymethylene polyphenylisocyanate (PAPI) was treated with water to initiate polymerization of PAPI. This increased the viscosity of the polymer to a point where it could be spread on the surface of the bars. The polymerization was brought about by water reacting with isocyanate groups to form amino groups which react with isocyanate groups on other PAPI molecules to increase the chain length and cross-linking. Then β-galactosidase powder was applied to the polymer on the bar, and the bar was immersed in 0.1 M phosphate buffer to couple the enzyme as shown in Figure 8. In phosphate buffer, the phosphate anion could react with isocyanate groups to form a mixed anhydride which would react with amino groups on the enzyme to produce urea bonds.

The immobilized β-galactosidase was very stable during repeated use with lactose as substrate and during long-term storage as shown in Table III. The first bar was used 43 times over an 85-day period with only a slight decrease in activity. The last two bars illustrate the good storage stability of the enzyme at 4°.

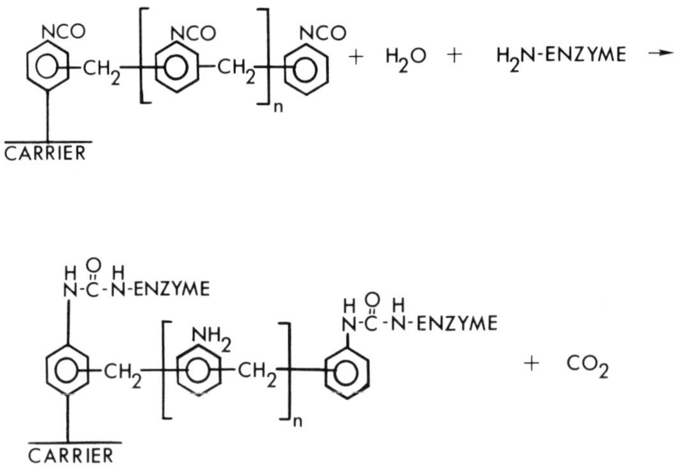

Figure 8. Reactions involved in polymerization of PAPI and coupling enzyme to polymer.

Table III. Stability of Three Immobilized β-Galactosidase Preparations after Cold Storage at 4°

Immobilized β-Galactosidase Preparation	Time days	No. 15-min Assays at 37°	Original Activity Remaining (%)
BAR-21	0	3	100
	29	21	86.1
	79	15	86.1
	85	4	86.1
BAR-19	0	3	100
	97	3	101
BAR-22	0	3	100
	97	3	95.4

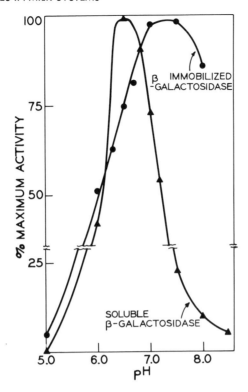

Figure 9. Hydrogen ion profile of lactose hydrolysis by native and immobilized β-galactosidase.

The shift in the pH optimum of the enzyme, shown in Figure 9, indicates that the matrix is negatively charged. However, the polymer surface should have a positive charge from protonation of the primary amines formed during prepolymerization of the polymer and during coupling of the enzyme. The negative charge could arise from β-galactosidase (active and denatured) and other proteins on the polymer surface.

Goodness of fit of the double reciprocal plot in Figure 10 indicates that this immobilized enzyme should have application for analytical purposes. Presumably the immobilization techniques used result in the enzyme being coupled to the polymer surface thereby reducing diffusion limitation. The Km of the immobilized enzyme at pH 7.3 was higher than the native enzyme (21.0 vs. 13.1). The Km of the bound enzyme at pH 6.5 was 22.1. Thus lactose could be estimated with this immobilized β-galactosidase in the range of 2 mM (0.1 Km).

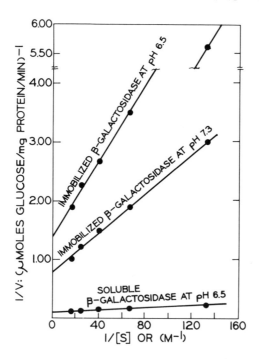

Figure 10. Lineweaver-Burk Plots comparing the activities of immobilized β-galactosidase with native β-galactosidase.

ACKNOWLEDGEMENTS

    Research was supported by the College of Agricultural and Life Sciences, University of Wisconsin, Madison; by the Cooperative Research Service, U. S. Department of Agriculture; by Public Health Service Grant No. FD-00200; and by Dairy Research Inc.

REFERENCES

Agner, K., in "Structure and Function of Oxidation-Reduction Enzymes," Akeson, A., Ehrenberg, A., Ed., Pergamon Press, New York, N.Y., 1972, p. 329.
American Public Health Association, Standard Methods for the Examination of Dairy Products, 12th ed., The Association, New York, 1967.

Ashoor, S. H., Sair, R. A., Olson, N. F., Richardson, T., Biochim. Biophys. Acta 229, 423 (1971).

Balcom, J., Foulkes, P., Olson, N. F., Richardson, T., Proc. Biochem. 6(8), 42 (1971).

Cuatrecasas, P., Wilchek, M., Anfinsen, C. B., Proc. Nat. Acad. Sci. U.S. 61, 636 (1968).

Dolgikh, T. V., Surovtsev, V. I., Kozlov, L. V., Antonov, V. K., Ginodman, L. M., Zvyatintsev, V. I., Prikl. Biokhim. Mikrobiol. 7, 686 (1971).

Dose, Z., Zaki, L., Z. Naturforsch. 26b, 144 (1971).

Ernstrom, C. A., "Rennin Action and Cheese Chemistry" in Fundamentals of Dairy Chemistry, Webb, B. H., Johnson, A. H., Ed., AVI Publishing Co., Westport, Conn., 1965, p. 590.

Ferrier, L. K., Richardson, T., Olson, N. F., Enzymol. 42, 273 (1972).

Ferrier, L. K., Richardson, T., Olson, N. F., Hicks, C. L., J. Dairy Sci. 55, 726 (1972).

Garnier, J., Ribadeau-Dumas, B., J. Dairy Res., 37, 493 (1970).

George, P., Nature 160, 41 (1947).

Green, M. L., Crutchfield, G., Biochem. J. 115, 183 (1969).

Hogg, D. McC., Jago, G. R., Biochem. J. 117, 779 (1970).

Hustad, G. O., Richardson, T., Olson, N. F., J. Dairy Sci. 56, 1111 (1973a).

Hustad, G. O., Richardson, T., Olson, N. F., J. Dairy Sci. 56, 1118 (1973b).

Imamoto, Y., Tsunemitsu, A., Okuda, K., J. Dent. Res. 51, 877 (1972).

Kirchmeier, O., Kolloid-Z. 236, 137 (1970).

Klebanoff, S. J., J. Clin. Invest. 46, 1078 (1967a).

Klebanoff, S. J., J. Exp. Med. 126, 1063 (1967b).

Klebanoff, S. J., J. Bact. 95, 2131 (1968).

Klebanoff, S. J., J. Reticulo-endothel. Soc. 12, 170 (1972).

Klebanoff, S. J., Clem, W. H., Luebke, R. G., Biochim. Biophys. Acta 117, 63 (1966).

Klebanoff, S. J., Luebke, R. G., Proc. Soc. Exp. Biol. Med. 118, 483 (1965).

Larson, W. A., Olson, N. F., Lund, D. B., J. Dairy Sci. 53, 646 (1970).

Lehrer, R. I., J. Bact. 99, 361 (1969).

Line, W. F., Kwong, A., Weetal, H. H., Biochim. Biophys. Acta 242, 194 (1971).

McMeekin, T. L., Groves, M. L., "Physical Equilibria in Milk: Proteins," in Fundamentals of Dairy Chemistry, Webb, B. H., Johnson, A. H., Ed., AVI Publishing Co., Westport, Conn., 1965, p. 374.

Miller, H., Biochem. J. 68, 275 (1958).

Morgulis, S., Beber, M., Rabkin, I., J. Biol. Chem. 68, 521 (1926).

O'Neill, S. P., Biotechnol. Bioeng. 14, 20 (1972).

Parry, R. M., Jr., Carrol, R. J., <u>Biochim. Biophys. Acta</u> <u>198</u>, 138
    (1969).
Payens, T. A. J., <u>J. Dairy Sci.</u> <u>49</u>, 1317 (1966).
Quarne, E. L., Larson, W. L., Olson, N. F., <u>J. Dairy Sci.</u> <u>51</u>, 848
    (1968).
Ribadeau-Dumas, B., Garnier, J., <u>J. Dairy Res.</u> <u>37</u>, 269 (1970).
Rose, D., <u>Dairy Sci. Abstr.</u>, <u>31</u>, 171 (1969).
Rosoff, H. D., Cruess, W. V., <u>Food Res.</u> <u>14</u>, 283 (1949).
Simmons, S. R., Karnovsky, M. L., <u>J. Exp. Med.</u> <u>138</u>, 44 (1973).

PREPARATION AND APPLICATION OF IMMOBILIZED β-GALACTOSIDASE

OF SACCHAROMYCES LACTIS

J. H. Woychik, M. V. Wondolowski, and K. J. Dahl

Eastern Regional Research Center, Agricultural Research

Service, U.S. Department of Agriculture, Phila., Pa. 19118

A variety of dairy products has been prepared in which the lactose has been partially or completely hydrolyzed by soluble β-galactosidases added to milk or whey (Kosikowski and Wierzbicki, 1971; Wendorff et al., 1971; Woychik and Wondolowski, 1973). These products were generally of good quality and acceptability and established the feasibility of using enzymes to hydrolyze lactose. Due to the high cost of soluble β-galactosidase it would appear that use of the enzyme in an immobilized form could offer significant economic advantages in the production of low-lactose dairy products.

Although studies of immobilized β-galactosidases of Escherichia coli (Sharp et al., 1969), Saccharomyces lactis (Dahlqvist et al., 1973), and Aspergillus niger (Olson and Stanley, 1973; Woychik and Wondolowski, 1972, 1973) have been reported in the literature, evaluation of the bound enzymes for lactose hydrolysis in dairy products has been limited primarily to the A. niger β-galactosidase. This enzyme has a pH optimum of 4.0 which limits its efficient use to acid wheys (pH 4.5). Skim milk and sweet whey (pH 6.3-6.8) require the use of a galactosidase with a higher pH optimum. The recent commercial availability of the β-galactosidase of the yeast Saccharomyces lactis (pH optimum, 7.0) prompted us to evaluate it in an immobilized form for lactose hydrolysis near neutral pH's.

This report is concerned with the study of the properties of S. lactis β-galactosidase immobilized on porous glass beads and on a new support, comminuted hide collagen.

41

MATERIALS AND METHODS

Çorning's Controlled-Pore porous glass beads (mean pore size 2050 Å, 80-120 mesh) were purchased from Electro-Nucleonics, Inc. A partially purified β-galactosidase of S. lactis (Maxillact) was generously supplied by Enzyme Development Corporation. The fibrous collagen was prepared from comminuted split hides which were dry ground, in the presence of solid carbon dioxide, to pass a 2 mm screen and lyophylized. The comminuted hide collagen is a research product of the Eastern Regional Research Center, Philadelphia. Reconstituted skim milk was prepared from commercial skim milk powder; the sweet whey was obtained from the Dairy Products Laboratory, ERRC, Washington, D.C.

Preparation of Bound Enzyme

The bound enzyme was prepared by two methods. (1) The β-galactosidase was attached to aminoalkylated glass beads using the glutaraldehyde procedures reported previously (Robinson et al., 1971; Woychik and Wondolowski, 1973). The aminoalkyl glass was suspended in a cold 1% aqueous solution of glutaraldehyde for 30 min, rinsed with water, and then suspended in cold phosphate buffer (pH 7, 0.1 M phosphate containing 0.01 M magnesium chloride) containing the β-galactosidase. After 2 hr, the glass was washed with phosphate buffer containing 0.1 M sodium chloride until no soluble galactosidase activity could be eluted. (2) The enzyme was bound to collagen by glutaraldehyde cross-linking. The lyophilized, fibrous collagen was allowed to swell in phosphate-magnesium buffer containing galactosidase (10 ml/g collagen) for 3 hr at room temperature. Sufficient 25% glutaraldehyde was then added to give a final glutaraldehyde concentration of 0.5% and the cross-linking allowed to proceed for 15 min. The collagen-enzyme preparation was washed repeatedly as above until the supernatant was free of galactosidase activity. The amount of protein bound to the supports was estimated by dividing the number of units bound per g support by the specific activity of the soluble enzyme, assuming the bound enzyme retained 100% of its original activity. Using the above techniques, preparations were obtained containing 8-14 mg enzyme protein/g support. Protein was determined by the method of Lowry et al. (1951) using bovine serum albumin as a standard.

Enzyme Activity and Lactose Hydrolysis

Enzymatic activity was determined by measuring the amount of glucose released following incubation at room temperature (25°) of either free or bound β-galactosidase with lactose. Glucose was determined using the glucose oxidase procedure of Jasewicz and

Wasserman (1961). Aliquots of digests containing the free enzyme
were inactivated by the pH drop accompanying the addition of the
glucose oxidase reagent (2 M sodium acetate, pH 4.0) or by placing
in a boiling water bath for 3 min. Activites of the immobilized
enzymes were determined by pumping substrate in downward flow
through columns of bound enzyme or by incubation with a stirred
suspension. Appropriate aliquots were analyzed directly for glucose
content by the glucose oxidase procedure. A unit of activity is
defined as the amount of enzyme which liberates 1 μmole of glucose/
min at room temperature (25°) using 5% lactose as substrate in pH
7.0 phosphate buffer. Specific activity is expressed as units/mg
protein.

The kinetic studies were done with the soluble and immobilized
enzymes using comparable amounts of activity (0.5-1.0 lactose units)
as determined under the standard conditions. One lactose unit
was equivalent to 66 μg of enzyme and was used as the standard
weight for the kinetic comparisons. The data were treated by the
least squares analysis using the weighting methods of Wilkinson
(1961).

RESULTS AND DISCUSSION

Enzyme Immobilization

A wide variety of insoluble supports and methods of enzyme immobilization have been reported; however, no single method or
support has been universally accepted. Porous glass beads have
probably been the most widely used support and have indeed proven
extremely useful in laboratory studies. Collagen offers several
advantages over other supports (Venkatasubramanian et al., 1972)
and has been utilized in membrane form for the immobilization of a
variety of enzymes (Wang and Vieth, 1973). The excellent mechanical
strength and hydrophilicity of collagen led us to investigate
its potential as an enzyme support in a particulate form.

The lyophilized collagen fibers swell rapidly and take up
approximately 100% their weight of buffered enzyme solution. Although maximum buffer uptake required approximately 1 hr, the
fibers were normally swelled in buffered enzyme solution for 3 hr.
Protein-protein interactions can lead to non-covalent bonding of
enzyme to the collagen fibers; however, after 16 hr, less than 0.5
mg enzyme/g collagen was bound in this manner. Therefore, glutaraldehyde was used to form cross-links between the enzyme and collagen. Figure 1 shows that the amount of enzyme bound in the
presence of glutaraldehyde depended on the protein concentration,
with a maximum binding of 8 mg/g collagen. Collagen preparations
containing 8 mg enzyme/g were used in the subsequent experiments.
Comparative experiments were done with the glass-bound galactosidase
having similar levels of enzymatic activity.

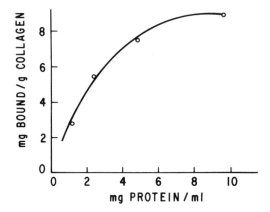

Figure 1.   Influence of β-galactosidase concentration on the
amount of enzyme bound to collagen.

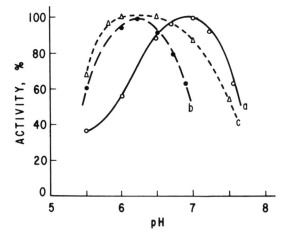

Figure 2.   Influence of pH on the activity of S. lactis β-galactosidase in free solution (-o-o-), bound to glass (--Δ--Δ--)
and bound to collagen (-●-●-).

Characterization of Bound β-Galactosidase

The pH activity curves were obtained for the free and bound enzymes at room temperature, using 5% lactose (0.146 M) in 0.1 M phosphate-0.01 M magnesium chloride buffers. Activities of the bound enzymes were determined in column operations. These results are presented in Figure 2. The maximum activity for the free enzyme (solid curve) occurred at pH 7.0, whereas the optima for the glass and collagen bound galactosidases occurred at pH 6.3-6.5. As noted in Figure 2, the curve for the enzyme bound to glass is quite broad in comparison to the other curves. The shifts in the pH optima can be attributed to either microenvironmental effects or to changes in the enzyme caused by the reaction with glutaraldehyde.

The effect of substrate concentration on the rate of lactose hydrolysis by the free and immobilized galactosidases is shown in Figure 3. The plots reflect the similarity of behavior of the free and bound enzymes. The values for K_M and V calculated from this data are presented in Table I. The slightly higher K_M values determined for the immobilized enzyme can be attributed

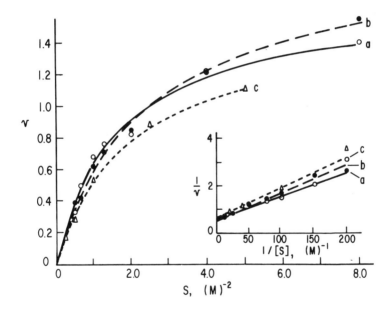

Figure 3. V versus S and Lineweaver-Burk plot showing the effect of substrate concentration on the hydrolysis of lactose by soluble β-galactosidase (a) and by β-galactosidase bound to collagen. (b) and bound to glass (c). The bound enzymes were analyzed in stirred reactors. Velocity is expressed as μmoles glucose released/min/standard amount of enzyme.

to steric factors resulting from covalent bonding to the supports
which may cause decreased binding of lactose to the active site.
The maximum velocities were comparable for the free and bound
enzymes.

TABLE I
Values of K_M and V for the Hydrolysis
of Lactose by Soluble and Immobilized
β-Galactosidase

	K_M(M)	V^a
Soluble	0.016	1.66
Glass-bound	0.019	1.60
Collagen-bound	0.022	1.92

[a]μmoles/min/standard amount of enzyme.

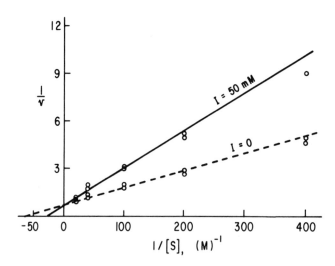

Figure 4.   Effect of galactose on the hydrolysis of lactose by
            soluble β-galactosidase in the presence of varying sub-
            strate concentrations. Velocity is expressed as μmoles
            glucose released/min/standard amount of enzyme.

With regard to the hydrolysis products, only galactose produced competitive inhibition; this is shown in the Lineweaver-Burk plot in Figure 4. The calculated K_i for galactose was 0.042 M.

The β-galactosidase of S. lactis is labile to temperatures above 30°; the stability was not materially improved by immobilization. Extended column operation of the bound enzymes was therefore limited to room temperatures. Neither the glass-bound nor collagen-bound galactosidases showed any loss of activity after 5 days of continual hydrolysis of 5% lactose at pH 7.0. After 3-4 days operation at room temperature, bacterial growth began to impair flow-rates. Although large amounts of bacteria could be removed from the supports by batch washing procedures, sufficient bacteria remained adsorbed and caused renewed plugging and channeling after 10-20 hr operation. A number of bactericides were studied to determine their potential use as column sanitizing agents. Hydrogen peroxide, benzoic acid, phenol, and iodofors all caused either partial or total inactivation of the galactosidase after 30 min exposure to working concentrations of these reagents. Although column operations at low temperatures (5-10°) reduced bacterial growth, column plugging still resulted after several days operation.

Application of Immobilized β-Galactosidase

The activity of the β-galactosidase toward lactose was influenced by the non-lactose solids in sweet whey and skim milk. At comparable pH's and lactose concentrations, the activity was decreased approximately 10% in sweet whey and 13% in skim milk when compared with the activity in buffered lactose solutions. These reductions in activity were also observed with the A. niger β-galactosidase (Woychik et al., 1973).

TABLE II
Hydrolysis of 5% Lactose[a] by
Collagen-Bound β-Galactosidase

Flow-Rate, l./hr	% Hydrolysis
2.3	93
3.6	84
4.4	80
5.2	76

[a] in 0.1 M phosphate, pH 6.7 containing 0.01 M magnesium chloride.

Some results obtained with a 70 g collagen-galactosidase column (4.5 x 35 cm) are presented in Table II. The amount of lactose hydrolyzed at pH 6.7 decreased with increased flow-rates and indicates that 75% or more hydrolysis is obtained at flow-rates up to 5 l./hr. A large column containing 375 g of collagen-bound galactosidase was operated at room temperature for the hydrolysis of lactose in sweet whey. Better than 85% hydrolysis was obtained at flow-rates up to 18 l./hr.

Although the collagen columns could be operated in downward flow for several days at slow flow-rates, rapid column packing resulted from flow-rates greater than 1 l./hr. Upward flow is therefore recommended for all column operations with collagen.

SUMMARY

The β-galactosidase of S. lactis has been immobilized to porous glass beads and to a new support, comminuted collagen. Collagen has been demonstrated to be comparable to other supports currently in use, and because of its hydrophilicity, it may facilitate substrate diffusibility in aqueous systems. The new properties conferred on the collagen by the reaction with glutaraldehyde make the tanned fibers quite suitable for batch or column operations.

The S. lactis β-galactosidase appears to have the necessary characteristics for successful applications to the hydrolysis of lactose in a variety of dairy products. The enzymic properties remain essentially unaltered after coupling with glutaraldehyde to either glass or collagen. This is reflected in the comparable kinetic data for the soluble and the immobilized forms.

The problem of bacterial growth associated with rich nutrient fluids, such as milk and whey, remains the major obstacle to commercial adaptation of an immobilized β-galactosidase system. The bacteria adsorbed to the insoluble supports, and their associated growth, result in column-plugging and in a generally unsanitary situation. There remain to be developed adequate bactericides or bacteriostats which will permit routine sanitization of immobilized enzyme reactors.

REFERENCES

Dahlqvist, A., Mattiason, B. and Mosbach, K. (1973). Biotechnol. Bioeng. 15:395.
Kosikowski, F.V. and Wierzbicki, L.E. (1971). J. Dairy Sci. 54:764.
Lowry, O.H., Rosebrough, N.J., Farr, A.L., and Randall, R.J. (1951). J. Biol. Chem. 193:265.

Olson, A.C. and Stanley, W.L.(1973). J. Agr. Food Chem. 21:440.

Sharp, A.K., Kay, G., and Lilly, M.D.(1969). Biotechnol. Bioeng. 11: 363.

Venkatasubramanian, K., Vieth, W.R., and Wang, S.S.(1972). J. Ferment. Technol. 50:600.

Wang, S.S. and Vieth, W.R.(1973). Biotechnol. Bioeng. 15:93.

Wendorff, W.L., Amundson, C.H., Olson, N.F., and Garver, J.C.(1971). J. Milk Food Technol. 34:294.

Wilkinson, G.N.(1961). Biochem. J. 80:324.

Woychik, J.H. and Wondolowski, M.V.(1972). Biochim. Biophys. Acta 289:347.

Woychik, J.H. and Wondolowski, M.V.(1973). J. Milk Food Technol. 36:31.

THE USE OF TANNIC ACID AND PHENOL-FORMALDEHYDE RESINS WITH GLUTARALDEHYDE TO IMMOBILIZE ENZYMES

Alfred C. Olson and William L. Stanley

Western Regional Research Center, Agricultural Research

Service, U.S. Department of Agriculture, Berkeley, CA 94710

Our investigations with immobilized enzymes have been guided by how they may assist in finding out more about enzymic phenomena and how they could ultimately be utilized in processes and analyses, particularly those related to agriculture and food. In the latter area economic as well as scientific factors play a very important role. A complete evaluation of all of these factors is beyond the scope of this paper. However, we have considered many of these factors in both the choice of the systems for immobilization and the applications we have elected to study. The criteria we used in selecting methods for enzyme immobilization included: successful achievement of stable enzyme immobilization; an acceptable degree of retention of enzyme activity; good pH and temperature characteristics; simplicity of the method; use of impure or inexpensive enzyme preparations; a potential for large volume application; and low cost.

Our investigations involving immobilized enzymes may be divided into three parts. First, we have studied the reaction of glutaraldehyde with enzymes and other proteins (DeJong et al., 1967; Jansen and Olson, 1969; Jansen et al., 1971; Tomimatsu et al., 1971; Gaffield et al., 1973). Second, in an attempt to apply the results obtained with glutaraldehyde we studied the immobilization of enzymes with tannic acid and glutaraldehyde (Stanley and Olson, 1973). Third, we have studied the immobilization of enzymes with phenol-formaldehyde resins and glutaraldehyde (Olson and Stanley, 1973).

In considering areas of applicability our attention was directed to the problem of the hydrolysis of lactose in milk and milk products using an immobilized lactase (β-galactosidase, EC 3.2.1.23). This problem has received a great deal of publicity in the past few years because many adults are unable to hydrolyze and utilize this

sugar due to the absence or low levels of intestinal lactase. The
ability to pre-hydrolyze lactose for those people unable to digest
it may then be a desirable objective. Lactases from several sources
exhibiting different enzyme characteristics have been immobilized by
several different procedures (Sharp et al., 1969; Dahlqvist et al.,
1973; Wierzbicki and Edwards, 1973; Woychik and Wondolowski, 1972,
1973; Woychik et al., 1974; Bernath and Vieth, 1974; Olson and
Stanley, 1973). In our investigation we have used a lactase from
Aspergillus niger with a pH optimum of about 4 (Lactase LP obtained
from Wallerstein Co.).

In this review we have summarized and correlated our observa-
tions on the reaction between glutaraldehyde and enzymes, the insolubilization of enzymes with tannic acid and glutaraldehyde and the
use of phenol-formaldehyde resins and glutaraldehyde to immobilize
enzymes, particularly a lactase from Aspergillus niger.

THE REACTION BETWEEN GLUTARALDEHYDE AND ENZYMES

Glutaraldehyde plays an important role in many methods for
enzyme immobilization. For this reason it appeared desirable to
consider the reaction between this difunctional aldehyde and enzymes.

From studies of the ^{1}H n.m.r. spectra of solutions of glutaraldehyde in deuterium oxide, Hardy et al. (1972) concluded that at
room temperature the major component in solution is a cyclic monohydrate with lesser amounts of free dialdehyde and acyclic dihydrate
and acyclic monohydrate. Commercial solutions of glutaraldehyde
may also contain significant quantities of α,β-unsaturated aldehydes
derived from the aldol condensation of glutaraldehyde as well as
other polymers and impurities such as acrolein and glutaric acid.
We have found that freshly distilled solutions of glutaraldehyde
reconstituted in water can be stored for several months in the cold
with little change occurring.

Hopwood (1972) has reviewed the theoretical and practical
aspects of glutaraldehyde fixation for studies of cellular ultra
structure including the reaction of glutaraldehyde with proteins.
An amino acid analysis of the product after the reaction of proteins
with glutaraldehyde shows only a change in the lysine values, which
can decrease to about 50-60% of the original lysine level for the
protein under investigation. The reaction is apparently progressive
with time and probably depends on the availability of the ε-amino
groups. Jansen et al. (1971) found that the pH optimum for the
most rapid insolubilization of proteins was near their isoelectric
point, though there were exceptions. The isoelectric point and
the pH for most rapid insolubilization of several proteins were:
bovine serum albumin, 4.7 and 4.8; soybean trypsin inhibitor, 4.6
and 4.8; α-chymotrypsin, 8.6 and 6.2; chymotrypsinogen-A, 9.5 and
8.2; papain, 8.75 and 8.6; and lysozyme, 11.0 and 10.5. With the

exception of α-chymotrypsin and its zymogen the correlation is good and indicates that protein charge is important for the intermolecular cross-linking required for insolubilization and that the optimum pH for intermolecular cross-linking is at the protein isoelectric point where repulsive charges between protein molecules are minimal.

In the case of the exception, α-chymotrypsin and its zymogen, the pH optima for insolubilization are lower than the respective isoelectric points of the native proteins. The suggested explanation for this is that an acid shift (from 10 to about 8.5) in the pKa of the ε-amino groups of glutaraldehyde-modified lysines in these proteins results in a shift in the isoelectric point from 8.2 to 6.2 and 9.5 to 8.2 for α-chymotrypsin and its zymogen respectively. For proteins with low (bovine serum albumin and soybean trypsin inhibitor) or high (lysozyme) isoelectric points, the glutaraldehyde-modified lysines are, respectively, either fully protonated or fully deprotonated at the protein isoelectric point so that the isoelectric point is not affected by the change in lysine pKa. This explanation, however, leaves the results for papain anomalous, since a similar shift for the same reason might have been expected.

This work was extended to investigating two proteins with different pH optima for glutaraldehyde insolubilization, α-chymotrypsin at pH 6.2 and bovine serum albumin at pH 4.7 (Jansen et al., 1971). The pH for concurrent complete insolubilization of both proteins was found to be 5.4, intermediate between the pH values for the individual proteins. Bovine serum albumin could not be insolubilized with glutaraldehyde at pH 6.2. In addition, it interfered with the complete precipitation of α-chymotrypsin at this pH since protein was only partially precipitated from a mixture of 100 mg of α-chymotrypsin and 100 mg of bovine serum albumin in 100 ml of 0.1 M phosphate buffer at pH 6.2 and 1.2% glutaraldehyde after 3 days at 25°. When treated by themselves under these same conditions α-chymotrypsin was completely precipitated while none of the bovine serum albumin was precipitated.

There was a loss of 50-70% of enzyme esterase activity during the initial reaction of papain and α-chymotrypsin with glutaraldehyde (Jansen et al., 1971). This loss in activity occurred before the enzyme was precipitated. About 15% of the initial esterase activity and 0.5% of the proteinase activity remained after insolubilization. Precipitation per se did not result in another large loss in enzyme activity. The loss in α-chymotrypsin esterase activity when it was coinsolubilized with bovine serum albumin was about the same as that of individually insolubilized α-chymotrypsin.

Insolubilization of α-chymotrypsin was found to be less rapid the higher the ionic strength of the reaction mixture (Tomimatsu et al., 1971). Increasing the ionic strength decreased the attractive forces between specific charged groups of the enzyme molecules resulting in a slower rate of insolubilization.

The formation of the cross-links between glutaraldehyde and
α-chymotrypsin was studied by Tomimatsu et al. (1971) using light
scattering measurements. The change in 90° scattering with time
showed that scattering increased rapidly at first and then slowed
down. This initial rapid increase in light scattering is attributed to intermolecular cross-linking with the reactive ε-amino
groups of lysine residues which are on or near the surface of the
molecule. At longer times there was a second rapid increase in
scattering that was linear with respect to time. This is the
result of linear polymerization of the polymers formed in the first
step by cross-linking to form larger polymers. The molecular
weights of these large soluble polymers range up to 220 X 10^6.
From molecular weight and radius of gyration data the shape of
these polymers appears to be branched flexible coils. As already
noted, there was an initial 50-70% loss in α-chymotrypsin esterase
activity on reaction with glutaraldehyde, which occurs during the
formation of the relatively small soluble polymers. This suggests
that reaction of glutaraldehyde at lysine ε-amino groups on or near
the surface of the molecule is the primary cause for the loss in
activity and not intermolecular cross-link formation.

Glutaraldehyde insolubilized α-chymotrypsin was not solubilized
by treatment with 6 M urea, alkali at pH 9.2 or acid at pH 3.5
(Jansen et al., 1971) showing that the insoluble derivative itself
is a chemically cross-linked product rather than an aggregate of
large, soluble, cross-linked polymers of the enzyme.

From the foregoing it is apparent that the reaction between
glutaraldehyde and proteins is a complex one which is far from
completely understood at this time. The physical form of glutaraldehyde insolubilized enzymes is noncrystalline. The particles tend
to clump and pack together and when placed in a column seriously
restrict passage of any liquid. For this reason when glutaraldehyde
is used as part of an immobilizing system there is usually another
insoluble phase upon which the glutaraldehyde insolubilized enzyme
is deposited or connected. Collagen (Bernath and Vieth, 1974;
Woychick et al, 1974) and nylon precipitated from formic acid
(Reynolds, 1974) are two examples of this other phase as are the
tannic acid and phenol-formaldehyde resins to be considered in the
next section. The addition of other components only increases the
complexity of the glutaraldehyde-enzyme reaction and at present
makes it more difficult to understand what reactions are taking
place.

INSOLUBILIZATION OF ENZYMES WITH TANNIC ACID AND GLUTARALDEHYDE

Negoro (1970) reported that invertase could be precipitated
with tannic acid and the precipitate mixed with filter aid and

packed in a column for the continuous hydrolysis of sucrose. In
applying this procedure to a β-galactosidase (lactase) from Aspergillus niger we found that while it was indeed possible to obtain
an active insoluble tannic acid-enzyme complex, on successive
washings of the complex to remove soluble enzyme the complex broke
down and essentially resolubilized. This difficulty was overcome
by treating the enzyme-tannic acid complex with glutaraldehyde
(Stanley and Olson, 1973). In a typical experiment 100 mg of lactase was dissolved in 5 ml of distilled water and a 10% solution
of tannic acid previously adjusted to pH 5.0 with sodium hydroxide
was added followed by 0.2 ml of a 25% solution of glutaraldehyde
(from Union Carbide). Precipitation was allowed to proceed at 5°
for 16 hours after which the immobilized lactase was collected by
centrifugation. Soluble enzyme was removed by repeated resuspension
of the solid in 10-ml portions of water and recentrifugation. Under
those conditions the insolubilized lactase did not redissolve.
Enzyme activity was measured using a 3% lactose solution and following the initial formation of glucose by the glucose oxidase -
chromagen procedure (Glucostat procedure by Worthington). Under
these conditions there was a 60% retention of the original enzyme
activity in the tannic acid-glutaraldehyde enzyme complex.

Other enzymes could also be insolubilized with tannic acid and
glutaraldehyde with initial retention of significant enzyme activity. Two examples of these with the % retention of activity and
substrate used are invertase (55%, sucrose) and glucoamylase (15%,
Lintner's starch).

The tannic acid-glutaraldehyde insolubilized lactase was too
amorphous and packed too tightly in a column to be useful as such
in continuous flow operation. It could, however, be mixed with
"Celite," a diatomaceous filter aid, in the ratio of 1 part insolubilized enzyme to about 2 parts "Celite," and successfully used in
column operation. The insolubilization of the lactase could also
be carried out in the presence of the "Celite" in which case there
was a more even distribution of the enzyme on the inert support.
When such a preparation of insolubilized lactase was packed into a
jacketed column 8 mm in diameter x 2 cm long over 90% hydrolysis of
a 3% lactose solution at pH 4.5, 45° and 0.6 ml/min was possible.
Such a column was operated continuously for over 72 hours with no
detectable loss in ability to hydrolyze lactose. While the level
of activity could be maintained for several days the difficulty
with the system as described is that the support packed together
causing the back pressure to build up on the column. Upflow operation was equally difficult since the particles were so fine and of
such a density that they moved up with the substrate and packed
against the upper column support.

The concept of using a phenol based solid support on which to
immobilize enzymes may have some relation to proteins found

complexed with tannins in nature. In order to pursue this idea
further we chose to investigate the use of phenol-formaldehyde
resins, phenolic polymers with better column characteristics than
the tannic acid complexes, as the basis for an immobilized enzyme
system.

INSOLUBILIZATION OF ENZYMES WITH PHENOL-FORMALDEHYDE RESINS AND GLUTARALDEHYDE

Phenol-formaldehyde resins have been available commercially for
a number of years as adsorbents for proteins and colored materials
from aqueous solutions. We have examined two such resins which we
obtained from Diamond Shamrock Chemical Company as Duolite S-30 and
Duolite Enzyme Support (Olson and Stanley, 1973; Olson and Stanley,
1973 patent; Olson and Stanley, 1974). The following account is a
summary of this information correlated with other related work in
our laboratory.

The resins are granular in the range 10-50 mesh with the
angular particles measuring from about 1.2 mm down to 0.1 mm. The
particles are relatively homogenous, translucent, amber, or reddish-
brown in appearance. They are quite porous as shown in the scanning
electron micrographs in Figure 1. The pictures show the fracture
face of a resin particle after treatment with lactase and glutaraldehyde. There were no observable differences between pictures of
Duolite S-30 and Duolite Enzyme Support, either with or without
enzyme and glutaraldehyde. Failure to detect the bound enzyme is
probably due to the fact that there is only 2-4% enzyme on the
resin, with no specific marker to locate protein by the techniques
employed. From these pictures the resin appears to have a compacted
granular texture. The smallest pore spaces between granules appear
to be on the order of 300 A° wide, close to the limit of resolution
of the scanning electron microscope. Smaller spaces might be seen
if the resolution of the instrument was greater.

Lactase and other proteins are readily adsorbed by Duolite
S-30 to the extent of 2-3 mg protein per gram of wet drained resin.
(Wet drained resin contains about 60% water.) Adsorption of lactase from aqueous solution by the resin was usually complete within
1-2 hours. Subsequent treatment of the adsorbed lactase with 1-3%
glutaraldehyde bound the enzyme firmly to the resin. The specific
activity of lactase adsorbed and coupled to Duolite S-30 was
80 μmoles of lactose hydrolyzed/min/mg enzyme or 200 μmoles lactose
hydrolyzed/min/g drained immobilized resin-enzyme. This activity
was determined by measuring the initial rate of glucose produced
when a 1.0-g portion of immobilized lactase was incubated at 45° in a
250-ml flask with 50 ml of 0.40 M lactose, 0.10 M in sodium acetate,
pH 4.0, with vigorous agitation. Considerable latitude in reaction
sequence was possible, including adding glutaraldehyde to the resin

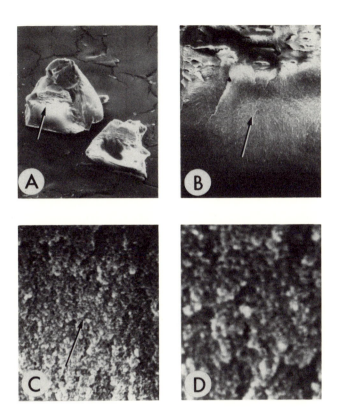

Figure 1. Scanning electron micrographs of Duolite phenol-formaldehyde resin after treatment with lactase and glutaraldehyde.
Particles were fractured and then mounted on double sticky Scotch
tape on glass slides and thinly coated with vacuum evaporated gold
by rotary deposition.

A. Two halves of a freshly fractured resin granule showing the
    fracture faces.   65X
B. Higher magnification of A in vicinity of arrow tip.   650X
C. Higher magnification of B in vicinity of arrow tip.   6,500X
D. Higher magnification of C in vicinity of arrow tip.   20,000X

(Micrographs were provided by Drs. R. S. Thomas and F. T. Jones
of this laboratory.)

first, followed by the enzyme, to adding the enzyme first, followed
by glutaraldehyde after 5 minutes up to 16 hours. Immobilization
was successful in the temperature range 5-20° and could be carried
out batchwise or with the resin packed in a column.

Duolite Enzyme Support is a phenol-formaldehyde resin that
has been sized and washed with acid, base and water. This resin
adsorbed 10-11 mg of lactase/g wet resin, considerably more than
the Duolite S-30. The specific activity of this immobilized lactase was correspondingly higher than that prepared from the S-30
at 500 μmoles lactose hydrolyzed/min/g of drained resin enzyme complex. This improvement in support loading and resultant specific
activity is not related to particle size and therefore may be due
to differences in available resin surface area for enzyme binding.
Identification of the factors responsible for this improvement and
their further application in resin preparation could result in even
higher loading capacities and specific activities.

The following enzyme performance characteristics of the immobilized lactase on either Duolite resin were similar. Activity
increased as the pH was decreased to 4.0 and the temperature was
increased to 60°. At temperatures of 50° and above for several
hours there was significant loss in activity. At 45° for 4 weeks
in the presence of lactose there was no measurable loss in activity.
All of the enzyme activity was lost on drying the immobilized lactase and 70% was lost on freezing. No loss was observed on storing
the immobilized lactase in the cold for 6 months.

In column operation performance of lactase on Duolite Enzyme
Support was considerably better than on Duolite S-30. Residence
time vs per cent hydrolysis for 0.40 M lactose over 7.0 g of the
former is shown in Figure 2. Since there was no change in this
curve between upflow or downflow operation or column diameters of
1.2 cm or 2.5 cm, it was concluded that within these parameters
performance is a function of residence time and is controlled by
support mass transfer or kinetic considerations rather than column
or reactor configuration.

Increasing the flow rate for a given column decreases the per
cent lactose hydrolyzed for any given original lactose concentration as shown by the data in Table I. For a single pass operation
to give a product that is 85-95% in hydrolyzed lactose the optimum
original lactose concentration for maximum hydrolyzed lactose per
hour is 6.8%. Increasing the desired level of hydrolysis decreases
the amount that can be hydrolyzed per hour. The higher the desired
level of conversion, the greater the effect of the original lactose
concentration on the amount of hydrolyzed lactose that can be
produced per hour.

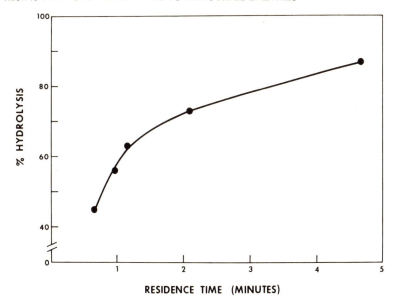

Figure 2. Effect of residence time on percent hydrolysis for lactase immobilized on Duolite Enzyme Support. For 7.0 g immobilized
lactase (wet weight) in a 1.2 or 2.5 cm diameter column operated
at 45°, with 0.40 M lactose, 0.10 M in sodium acetate, pH 4.0.

TABLE I

Grams of Lactose Hydrolyzed vs Original Concentration and
Percent Hydrolysis[1]

Original lactose, %	Grams hydrolyzed/hour at –		
	85	90	95% Hydrolysis level
3.4	8.3 (288)	7.3 (242)	6.5 (200)
6.8	10.1 (175)	8.6 (140)	7.1 (110)
13.7	9.4 (80)	6.7 (55)	3.9 (30)
17.1	7.3 (50)	5.2 (34)	3.4 (15)

[1] For 7.0 g lactase on Duolite Enzyme Support, pH 4.0, 45°.
Numbers in parentheses are the flow rates in ml/hour for
the designated hydrolysis levels.

No problems were encountered when the small column system just described was scaled up over 500X. Then, in order to get some idea of projected plant size and assuming that direct scale-up is possible, the volumes of resin necessary to hydrolyze 100 kg of lactose/hour at different levels of hydrolysis and original lactose concentrations were calculated (Table II). Thus, for example, 90% hydrolysis of a 6.8 solution would require passage of 111 kg of lactose over a 119-liter column to give 100 kg of hydrolyzed lactose. The original lactose concentration that requires the smallest volume of resin to meet the requirements is 6.8% at all three levels of hydrolysis. A column measuring 39 cm in diameter and 100 cm in height would hold 119 liters of immobilized lactase and running 24 hours a day for 300 days a year would theoretically hydrolyze more than a million pounds of lactose.

The problem of keeping the immobilized enzyme system free of microbiological contamination is a formidable one. Our procedure with the lactase columns has been to rinse columns frequently with water or dilute hydrogen peroxide, particularly before shutting them down. When columns were not being used they were stored in the cold. In most instances this has been sufficient to keep microbiological contamination under control. The phenol-formalde-

TABLE II
Calculated Volume of Resin Necessary
to Hydrolyze 100 kg Lactose/Hour[1]

Original lactose, %	Volume of resin in liters required for		
	85	90	95% Hydrolysis
3.4	116	139	167
6.8	95	119	151
13.7	103	151	277
17.1	133	196	444

[1] At pH 4.0, 45°.

hyde resin itself and glutaraldehyde are both unfavorable substrates for micro-organisms, if not antimicrobial.

From the discussion of the reaction of enzymes with glutaraldehyde it is apparent that some reaction is occurring between glutaraldehyde and lysine ε-amino groups resulting in cross-linking of enzyme molecules. Glutaraldehyde-insolubilized α-chymotrypsin and other proteins were shown to be insoluble in 6 M urea, demonstrating the covalent nature of the reaction as well as its irreversibility. Thus, as one might expect, the protein fixed to resin with glutaraldehyde does not easily come off and cannot be resolubilized. Secondly, in spite of the failure to observe differences in the order in which enzymes and glutaraldehyde were added to the resin it would appear that this order could influence the nature of the final product. There may be a difference between cross-linking an enzyme after it is adsorbed to resin and depositing partially cross-linked enzyme on the resin where it could be further reacted with glutaraldehyde.

REFERENCES

Bernath, F. R., Vieth, W. R., This volume in chapter entitled, "Collagen as a Carrier for Enzymes" (1974).

Dahlqvist, A., Mattiason, B., Mosbach, K., Biotechnol. Bioeng. 15, 395 (1973).

DeJong, D. W., Olson, A. C., Jansen, E. F., Science 155, 1672 (1967).

Gaffield, W., Tomimatsu, Y., Olson, A. C., Jansen, E. F., Arch. Biochem. Biophys. 157, 405 (1973).

Hardy, P. M., Nicholls, A. C., Rydon, H. N., J. Chem. Soc. Perkin Trans., II 15, 2270 (1972).

Hopwood, D., Histochem. J. 4, 267 (1972).

Jansen, E. F., Olson, A. C., Arch. Biochem. Biophys. 129, 221 (1969).

Jansen, E. F., Tomimatsu, Y., Olson, A. C., Arch. Biochem. Biophys. 144, 394 (1971).

Negoro, H., Hakko Kogaku Zasshi 48, 689 (1970).

Olson, A. C., Stanley, W. L., J. Agr. Food Chem. 21, 440 (1973).

Olson, A. C., Stanley, W. L., U.S. Patent 3,767,531 (1973).

Olson, A. C., Stanley, W. L., in "Enzyme Engineering, Vol. 2."
    (Eds. E. K. Pye and L. B. Wingard, Jr.) Plenum Press, New York
    (1974).

Reynolds, J. H. This volume in chapter entitled, "The Uses of
    Precipitated Nylon as an Enzyme Support; an α-Galactosidase
    Reactor" (1974).

Sharp, A. K., Kay, G., Lilly, M. D., Biotechnol. Bioeng. 11, 363
    (1969).

Stanley, W. L., Olson, A. C., U.S. Patent 3,736,231 (1973).

Tomimatsu, T., Jansen, E. F., Gaffield, W., Olson, A. C., J.
    Colloid Interface Sci. 36, 51 (1971).

Wierzbicki, L. E., Edwards, V. H., Kosikowski, F. V., J. Food Sci.
    38, 1070 (1973).

Woychik, J. H., Wondolowski, M. V., Biochim. Biophys. Acta 289,
    347 (1972).

Woychik, J. H., Wondolowski, M. V., J. Milk Food Technol. 36, 31
    (1973).

Woychik, J. H., Wondolowski, M. V., Dahl, K. J. This volume in
    chapter entitled, "Preparation and Application of Immobilized
    β-Galactosidase of Saccharomyces Lactis," (1974).

THE USES OF PRECIPITATED NYLON AS AN ENZYME SUPPORT:

AN α-GALACTOSIDASE REACTOR

J. H. Reynolds

Monsanto Company

St. Louis, Missouri 63166

ABSTRACT

Nylon precipitated from formic acid produces a surface into which enzymes readily adsorb. They can be
immobilized on the surface by inter-enzyme crosslinking
with glutaraldehyde or dimethyladipimidate. Nylon has
been deposited onto the surfaces of porous polyethylene
disks which provide a very convenient rigid matrix for
a small enzyme reactor. On the other hand, the precipitated nylon does not require a support. It can be used
to fill large column reactors through which enzyme and
crosslinking agent are passed to make the immobilized
enzyme in situ. A new, neutral, thermostable α-
galactosidase of molecular weight 150,000 has been
immobilized by this latter technique for the purpose of
hydrolyzing the oligosaccharides in soy milks and beet
sugar molasses. In addition thermolysin, papain,
glucose isomerase and amyloglucosidase reactors have
been prepared and studied.

INTRODUCTION

The use of solid surfaces as a method of separating
reactants and products of chemical reactions has markedly
affected both synthetic and analytical biochemistry in
the last two decades (Stark, 1971). Examples are amino
acid analysis (Spackman et al., 1958), gel permeation
chromatography (Determann, 1968), the synthesis of pep-

tides and proteins (Marshall and Merrifield, 1971) and
affinity chromatography (Cuatrecasas and Anfinsen, 1971).

This list continues to grow. A relatively new
technique is solid phase radioimmunoassay (Catt, 1969).
The growth of primary animal cells in vitro has been
studied on many surfaces and a continuous reactor using
cells attached to hollow fibers has been described
(Knazeh, et al., 1972). A number of common organic
reagents have been immobilized. For example, Blossey
et al. (1973) have bound AlCl$_3$ to polystyrene and used
the product as an esterification catalyst. The immobilization of enzymes is just one of the many techniques
in which solid surfaces increase the facility of many
laborious tasks.

From our early work in the area of immobilized
enzymes, it became apparent that if they ever were to
be used on a large industrial scale, their supports
would require much different properties. We have developed several superior support systems. One of these
is the use of precipitated nylon. Its application to
food chemistry will be discussed.

EXPERIMENTAL

Materials

Reagent chemicals were obtained from Fisher
Scientific, enzymes from Worthington Biochemical and
substrates from Sigma. Molasses, soy milk and beer were
obtained from potential customers.

Methods

Precipitated nylon floc was prepared according to
Reynolds (1972) and the α-galactosidase reactor prepared according to Reynolds (1974). The papain, thermolysin and glucose isomerase reactors were prepared in
a similar manner.

Trypsin Disk

A polyethylene disk 38 millimeters in diameter and
1/8 inch thick was cut from a sheet of porous poly-

ethylene and was soaked in a 5 weight percent solution
of nylon in formic acid at 50° C. Excess polymer solution was removed from the outer surfaces of the disk,
and the disk was placed in a membrane holder. Distilled
water was then pumped through the disk to precipitate
the nylon on the surfaces and to remove the formic acid.
A 10^{-2} M triethanolamine buffer at pH 8.5 containing
1 mg/ml trypsin, 10^{-3} M benzamidine, and 10^{-1} M dimethyladipimidate was pumped repeatedly through the disk for
one-half hour. The disk was washed with tris buffer
(pH 8.0) containing 1 M potassium chloride.

The disk was assayed by recycling the substrate(BAEE)
solution through it rapidly ($\sim$100 ml/min). A spectrophotometer with a flow cell is connected to the effluent
side of the disk. At the outset of the assay, the tube
from a substrate reservoir is attached to the inlet side
of the reactor with the outlet tube from the spectrophotometer going to waste. The substrate is pumped
through the disk at such a high rate that the conversion
is virtually nil. When that condition is reached, the
outlet tube is put into the substrate reservoir and the
rate observed by the spectrophotometer recorder tracing.
The volume is measured and the enzyme units calculated
in the same manner as it is done for enzymes in solution.

RESULTS AND DISCUSSION

Enzyme Disks

One of the most important attributes of a solid
support is its resistance to compression under high flow
and back pressures. We found that rigid porous polyethylene disks, once coated with a polymer such as nylon
or polyacrylonitrile, would absorb proteins. These
could then be immobilized with a difunctional cross-
linking agent such as glutaraldehyde or a bisimidate
ester. Similar techniques have been reported by Haynes
and Walsh (1969) and Olson and Stanley (1973).

Very convenient small laboratory reactors can be
made in this manner. The disk can be mounted into an
appropriate disk holder, used, removed and stored in
buffer at 4° until reuse. The trypsin disk contained
about 1 mg of enzyme (10 mg/gram nylon). This size
disk will hydrolyze 75% of 10^{-3} M benzyol arginine ethyl-

ester at pH 8.0 with a retention time of 0.3 sec. The
immobilized trypsin also hydrolyzes casein and heat
denatured lysozyme, but not native lysozyme. The disk
was used and stored over a 3-month period without
activity loss. A number of other enzyme disks have been
prepared and studied (Reynolds, 1972). To increase
reactor size, many small disks can be stacked in a long
reactor tube or a large diameter disk can be used. Both
of these techniques are cumbersome and do not lend themselves to easy manufacture of large reactors.

Nylon Floc

In the case of nylon we found that we could do without the polyethylene disk and the precipitated polyamide
could be used directly to adsorb enzyme and make a plug
flow reactor. By pouring a formic acid solution of
nylon into a large excess of highly agitated water, a
spongy, flocculant material is formed. The surface area,
by BET measurement, is 7 m^2/gram after the nylon is
dried. However, this does not represent the surface onto which the enzymes are absorbed since drying reduces
the binding ability to practically nil. Scanning electron microscopy shows a highly irregular mass of nylon
microfibrils.

The enzymes are immobilized onto the surface of the
nylon by reacting them with a bifunctional crosslinking
agent such as glutaraldehyde or a bisimidate ester, e.g.
dimethyladipimidate. It is proposed that the enzyme is
adsorbed to the polymer surface and crosslinked to other
enzyme molecules rather than to the polymer surface. If
the nylon amino groups are first amidimated with methyl-
acetimidate, the same amount of active enzyme is b nd
to the nylon as in the experiments in which the nylon
amino groups were not blocked. In general about 5-10 mg
protein can be bound per gram of nylon floc.

This nylon material is extremely easy to make, is
cheap, is very convenient to work with, is truly insoluble, does not dry out rapidly, is hydrophilic, is
reasonably rigid and can contain reasonable amounts of
enzyme which does not leach. It is also non-biodegrad-
able and resistant to chemical attack, thus making it a
good support for enzyme reactors. It does compact under
high back pressures or when very viscous solutions are
used. At least a dozen different enzymes have been
attached to the precipitated nylon (Reynolds, 1972).

α-Galactosidase

Our goal was to build a reactor which would hydrolyze the oligosaccharides in soy milks and in beet sugar
molasses. Both solutions must be kept near pH 7. High
raffinose concentrations in beet sugar molasses seriously inhibits further sucrose recovery and the high stachyose concentration in soy is at least one of the factors
responsible for intestinal discomfort and flatulence.
No thermostable, neutral α-galactosidase being known to
us, we produced one from a B. stearothermophilus (Kuo
et al., 1974). The enzyme was harvested and partially
purified (Weeks and Johnson, 1974). The enzyme has a
molecular weight of about 150,000 and is similar to
other α-galactosidases (Dey and Pridham, 1971). It
hydrolyzes p-nitrophenyl-α-D-galactopyranoside, raffinose, stachyose and melibiose. It has no protease or
invertase activity. It is inhibited by raffinose and
the optimum raffinose concentration is 1.9%.

α-Galactosidase-nylon reactors with total volumes
from 60 ml to 7 liters have been made. The yield of
active enzyme was 95%. About 10 mg of protein was bound
per gram of nylon floc. The large reactor required
spacer plates to keep the nylon from compacting under
the high flow rate used (2.5 liters/min) with 20 psi
back pressure. The immobilized α-galactosidase has a
long life time and has been used continuously for one
month at room temperature, hydrolyzing 1.9% raffinose
with no loss of activity. The immobilized enzyme has
the same kinetic characteristics for raffinose as the
free enzyme; pH optimum (7.0), K_m (1.4 x 10^{-2} M), K_i
(1.0 x 10^{-1} M) and optimum substrate concentration
(3.67 x 10^{-2} M).

The free and immobilized α-galactosidase were found
to hydrolyze both the stachyose in soy milk and the raffinose in beet sugar molasses in batch tests. A 1/10
diluted molasses originally containing 16% raffinose was
run through a 60 ml α-galactosidase/nylon reactor at
25°. Immediate decay of activity was observed and insoluble material began to build up in the inlet side of
the reactor. Filtering of the molasses through Whatman
#1 filter paper and an 8 μm Millipore filter removed the
inhibitory materials from the diluted molasses. No
plugging of the reactor occurred over a 7-day, continuous
operating period at 25° C even though the nylon adsorbed
pigments from the substrate stream. Reactors were built
which would hydrolyze 95% of a 1.5% raffinose solution
with a one-hr retention time. If lower dilutions of

molasses were used, the viscosity was great enough to
compact the nylon so that the back pressure essentially
became infinite and the reactor no longer useful. Tests
were run using lower molasses dilutions at 40°. Under
these conditions, bacterial growth (L. mesenteroides)
was enormous. The reactor required 370 ppm formaldehyde
(a common bacteriostat used in the sugar beet industry)
to remain free of bacterial growth. This concentration
of formaldehyde reduces the enzyme activity, but to a
constant level.

As expected, pumping soy milk containing 2.5%
stachyose at 60 ml/hr reduced the activity of a 60 ml
reactor, but did not plug it nor change the back pressures
over a 24-hr period of constant operation. The insoluble
materials in the soy milk built up on the nylon and physically blocked the substrate from reacting the enzyme.
The activity of the reactor was reduced by 50% in the
24-hr period, but could be completely restored by washing
the nylon with tap water. Further tests are required to
determine the best reactor design and reaction conditions.

Thermolysin

A thermolysin (a neutral protease) reactor was made
by recirculating aqueous enzyme and glutaraldehyde through
a 1" x 17" nylon packed reactor for 20 min at pH 6.5.
The excess reagents were washed out with water. Free
thermolysin autolyzes rapidly, and the increase in its
temperature stability upon immobilization is dramatic.
In this case, immobilization causes a tenfold increase
in shelf life of the enzyme at 70° in the presence of a
1% casein solution. When used to hydrolyze a soy protein isolate (Promine D), an insoluble product of hydrolysis physically blocked the column operation and reduced
its activity to zero. Again reactor design and possibly
the use of one or more other proteases in conjunction
with the thermolysin could solve this problem.

We have built other enzyme reactors for food processing. Papain was simply absorbed onto nylon without
the use of a crosslinking agent. It was used in several
experiments which showed that papain immobilized in this
manner will chillproof beer. Further tests are required
to ascertain its usefulness in pilot plant operations.
In addition we have built an amyloglucosidase reactor
from a B. subtilis saccharifying amylase which will
continuously hydrolyze soluble starch to glucose. A
glucose isomerase reactor is presently undergoing
laboratory tests.

All of the reactors described here work well with no leaching under ideal conditions with clear buffered substrates. Some suffer from both physical stoppage and reversible loss of catalytic activity because of product/reactant adsorption; other kinds of reactor design need to be considered. In the case of simple fouling due to particulate matter insolubles, a reactor design described by Wang and Vieth (1973) may be helpful. This allows insolubles to pass through the reactor and yet allows contact of substrate and enzyme. Nylon sheets precipitated from formic acid could be used rather than the collagen or DEAE cellulose. In addition a simple, constantly stirred tank reactor might solve some of these problems.

ACKNOWLEDGEMENTS

I would like to acknowledge the assistance of the following colleagues: B. S. Wildi, J. H. Johnson and L. E. Weeks.

REFERENCES

Blossey, E.C., L.M. Turner and D.C. Neckers, Tetrahedron Letters, 1823 (1973).

Catt, K.J., Acta Endocrinol., 63, Suppl. 142, 242 (1969).

Cuatrecasas, P. and C.B. Anfinsen, in Methods in Enzymology, Vol. XXII, W.B. Jakoby, Ed., p. 345 (1971).

Determann, H., Gel Chromatography, Springer-Verlag, New York (1968).

Dey, P.M. and J.B. Pridham, Advances in Enzymology, A. Meister, Ed., Wiley, New York, p. 91 (1971).

Haynes, R. and K.A. Walsh, Biochem. Biophys. Res. Commun., 36, 235 (1969).

Knazeh, R.A., P.M. Gullino, P.O. Kohler and R.L. Dedrick, Science, 178, 65 (1972).

Kuo, M.J., J. Delente and R.J. O'Connor, Biotechnol. and Bioeng., submitted for publication.

Marshall, G.R. and R.B. Merrifield, in Biochemical Aspects of Reactions on Solid Supports, G.R. Stark, Ed., Academic Press, p. 111 (1971).

Olson, A.C. and W.L. Stanley, J. Agr. Food Chem., 21, 440 (1973).

Reynolds, J., Biotechnol. and Bioeng., in press (1974).

Reynolds, J., U.S. Patent 3,705,084, "Macroporous Enzyme Reactor," (1972).

Spackman, D.H., W.H. Stein and S. Moore, Anal. Chem., 30, 1190 (1958).

Stark, G.R., Ed., Biochemical Aspects of Reactions on Solid Supports, Academic Press, New York (1971).

Wang, S.S. and W.R. Vieth, Biotechnol. and Bioeng., 15, 93 (1973).

Weeks, L.E. and J.H. Johnson, Biotechnol. and Bioeng., submitted for publication (1974).

GLUCOSE ISOMERASE CELLS ENTRAPPED IN CELLULOSE ACETATES

M.J. Kolarik, B.J. Chen, A.H. Emery, Jr., and H.C. Lim

School of Chemical Engineering, Purdue University

West Lafayette, Indiana 47907

Summary

Entrapment of whole cells containing glucose isomerase in primary and secondary cellulose acetates was investigated. These enzyme-acetate complexes were formed into fibers and membranes to the extent of one gram of whole cells entrapped in two grams of acetate. With fibers measuring about 250 x 500 microns in cross section the rate of diffusion of substrate into the fiber limits the activity of the resulting cell-acetate complex to the extent of 17% of the original cell activity. On the other hand, with membranes measuring 10 - 20 microns thick the activity was as high as 57% of the original cell activity. However, cells leaked from these membranes. The leakage of cells from other membranes was reduced to about 5% by varying the solvent system. Permeability was also affected by the solvent system used.

Introduction

In the last few years tremendous progress has been made in immobilizing various soluble enzymes on inert supports by various methods, and for detail one may refer to various review papers, for example, Melrose (1). It has also become apparent that to obtain highly active enzyme-support complexes one needs to use relatively pure enzyme preparations. Since many potentially important enzymes

71

are intracellular they must be released from the cells, placed in solution and then purified before they can be recovered in relatively pure form. Consequently, the isolation and purification of intracellular enzymes are in general time-consuming and the yields are generally very poor due to the many steps involved and often due to poor stability of the soluble enzymes. All these factors lead to relatively expensive enzyme preparations which, in turn, have to be immobilized. This disadvantage can be overcome by an alternative route, i.e., immobilization of whole cells.

The first reported work dealing with immobilization of whole cells containing enzymes is due to Leuschner (2) who immobilized free enzymes and microorganisms containing enzymes using natural and synthetic polymeric materials. More recently Vieth et al. (3) used collagen as a host matrix to immobilize Streptomyces phaeochromogenes cells containing glucose isomerase. Dinelli (4) reported entrapping of enzymes within the pores of wet-spun synthetic fibers and showed good activity and stability.

Among the advantages of the entrapping method are 1) it is quite general, unlike methods such as covalent bonding which rely on specific functional groups on enzymes and supports, so that a wide range of enzymes or enzyme-bearing materials can be used; 2) relatively inexpensive forms of enzymes, such as crude enzyme preparations or whole cells containing enzymes, can be used; and 3) the operation is very simple, mild, and inexpensive. Thus, this method is industrially very attractive. Primary and secondary acetates meet all of the above advantages. An added advantage of this system is that not only can it be used to entrap enzyme(s) or enzyme-bearing material, but also any solid material can be coated with this enzyme-acetate complex. Entrapping of whole cells in these acetates is an approach that has not been thoroughly investigated, although it has received some cursory treatment (2,4).

Cellulose acetates have been used in the form of yarns for making clothing, in the form of film for various packaging, and also in the form of membrane for desalination of sea water and artificial kidneys. Thus, its acceptability in food and drug applications is quite good. There is also much information available, some of which may be useful.

We report here entrapping of whole cells containing glucose isomerase in primary and secondary cellulose acetates in the form of fibers and membranes. This method involves dissolving the acetates in a solvent or solvents, adding whole cells, and evaporating the solvent(s) by casting the mixture on a flat surface, or coagulating the mixture in fiber form in a solvent by injecting the mixture through a syringe.

MATERIALS AND METHODS

Cellulose Acetates

Cellulose acetates were purchased from Eastman Kodak. The products actually used were the triacetate, flake form, and a secondary acetate powder designated as E-398-3.

Whole Cells

Whole cells containing glucose isomerase were obtained from Novo Enzyme Corporation, designated as SP-92.

Assay of the Enzyme Activity

In all cases the enzymatic activity of a preparation was determined by measuring the rate of generation of glucose from 0.1M fructose at 60°C in the presence of $0.001\underline{M}$ C_0^{++} and $0.01\underline{M}$ Mg^{++}. Glucose was measured by the glucose oxidase-peroxidase method.

Buffers

In all cases the reactions were carried out in 0.01 $\underline{M}$ succinate buffer. With preparations involving cellulose triacetate as a support, reactions were run at pH 6.5 while pH 6.8 was used with the secondary acetate preparations.

Solvents

Reagent quality acetone, toluene, and methylene chloride were used in this work.

Preparation of Cellulose Triacetate-Cell Solution

In a beaker 10 g. of cellulose triacetate were dissolved in methylene chloride to bring the total volume to 100 mls. In a separate beaker 5 g. of whole cells were slurried in 15 mls. of water. The whole cell solution was then poured into the acetate solution which was being stirred vigorously to prepare an emulsion of the aqueous phase in the organic phase. After stirring for about a half hour, secondary solvents, if used, were then added and stirring continued for five additional minutes. This mixture was then cast into fibers and membranes.

Preparation of the Secondary Acetate-Cell Solutions

To break up aggregates of the whole cells it was necessary to tumble the dry cells with the secondary acetate powder. This was done overnight with a mixture consisting of 2 g. of acetate and 1 g. of whole cells. While agitating, one gram of this mixture was then sprinkled into 14 mls. of solvent to prepare the solution used for making membranes.

Preparation of Fibers

Fibers were formed only with cellulose triacetate (CTA) for this work. A 10 ml. hypodermic syringe was filled with the CTA-cell solution and then used to extrude the solution through a number 20 hypodermic needle, which had been cut blunt, into 100 mls. of toluene. The resulting fibers were removed from the toluene and air-dried before use. These fibers generally consisted of ribbons measuring about 250 x 500 microns.

Casting Membranes on Water

This method requires a water soluble solvent, hence it was used only with the secondary acetate which was dissolved in acetone. The cellulose acetate-whole cell solution was simply dripped onto water from a Pasteur pipette. On the water surface the acetate solution would spread out, then solidify to form membranes which were normally less than 10 microns thick.

Casting of Membranes on Glass

This method is applicable for any solvent system and was used with both types of acetates. The acetate-cell solution was simply poured onto a flat glass plate, spread into a thin layer, and then either allowed to air dry or was immersed in water to remove acetone solvent. The resulting membranes generally measured 10 - 20 microns thick.

Various cell-acetate preparations are summarized in Table 1.

RESULTS

Results with Fibers

For early fibers only CH_2Cl_2 was used as a solvent for cellulose triacetate. This material was packed into a column and assyed with 0.1 M fructose. Table 2 summarizes the percentage of the enzymatic activity (expressed as the percentage of free cell activity) which could be observed for two of these columns.

It was felt that diffusion of substrate into the fiber was responsible for the low activities observed. The continuity equation for diffusion through a solid coupled with reversible chemical reaction of Michaelis-Menton type kinetics and one dimensional plane geometry is:

$$D_{AS} \ \nabla^2 S = \frac{\frac{V_1}{K_S} S - \frac{V_2}{K_P} P}{1 + \frac{S}{K_S} + \frac{P}{K_P}} \tag{1}$$

Table 1

SUMMARY OF PREPARATIONS MADE

	Material	
Preparation	CTA	CDA*
Fibers	Yes	No
Membranes on Water	No	Yes
Membranes on Glass -- wet cast	No	Yes
Membranes on Glass -- air dried	Yes	Yes

*Cellulose diacetate (Eastman Kodak, E398-3)

Table 2

ACTIVITY OF CELLULOSE ACETATE FIBERS

mls CH_2Cl_2/10 g CTA	% Activity
100	2
150	7

$V_1 = V_{max}$ for substrate

$V_2 = V_{max}$ for product

K_S = Michaelis constant for substrate

K_P = Michaelis constant for product

D_{AS} = diffusivity of the substrate

D_{AP} = diffusivity of the product

S_0 = initial substrate concentration

P_0 = initial product concentration

Define:

$$\alpha_1 = \frac{V_1}{K_S} + \frac{V_2}{K_P}$$

$$\alpha_2 = \frac{D_{AS}}{D_{AP}} S_0 - \frac{V_2}{K_P} P_0$$

$$\alpha_3 = \frac{1}{K_S} + \frac{D_{AS}}{D_{AP}} \frac{1}{K_P}$$

$$\alpha_4 = 1 + \frac{P_0}{K_P} + \frac{D_{AS}}{D_{AP}} \frac{S_0}{K_P}$$

The equation was readily integrated once and the constant of integration evaluated to give

$$\frac{dS}{dy} = \left(\frac{2 \alpha_1}{D_{AS} \alpha_3} \right)^{1/2} \left(S - S_c + \frac{\alpha_2}{\alpha_1} - \frac{\alpha_1 \alpha_4}{\alpha_3^2} \ln \frac{\alpha_3 S + \alpha_4}{\alpha_3 S_c + \alpha_4} \right)^{1/2} \quad (2)$$

where

S_c is the concentration of substrate at the centerline.

The rate at which product is formed from a column for small conversions is given in Equation (3).

$$\text{rate} = A\ D_{AS}\ \frac{ds}{dy}\bigg|_{surface} \tag{3}$$

A is the surface area of the fiber in the column and the last term is obtained from Equation (2) by using S as the concentration at the surface of the solid fiber.

When diffusion is the limiting factor the concentration of the substrate, S_C, will reach the equilibrium value ($\approx$ 1/2 concentration of the feed for this reaction). For feed concentrations not small with respect to K_S, the logarithmic term will be small with respect to $S - S_C$ and the rate of production from such a column would be expected to be approximately proportional to the feed concentration to the one half power.

Various feed concentrations were pumped through the column and the results are given in Figure 1. The line drawn through the data was generated from Equation (3) after the data point for the highest concentration was used to calculate a value for the terms independent of concentration. When the obtained data were treated as if there were little diffusional resistance and K_m evaluated, the resulting value of K_m was approximately nine times larger than for the free cell. We do not expect K_m to increase much by entrapment. Armed with this information we started looking for a way to increase the permeability of the cellulose acetate fiber.

Much work has been done to evaluate permeabilities of membranes made from other common cellulose derivatives. All appear to be rather impermeable in that they filter out a large portion of the salt in a salt solution, for example. We investigated the effect of adding various chemicals to the methylene dichloride solvent: alcohol, ketone, ester, near-solvent, non-solvent. Two levels of secondary solvent were used: to 100 mls. of CTA-CH_2Cl_2 solution either 10 mls. or 100 mls. of the second solvent were added. Table 3 summarizes the results from this work, showing the percentages of added enzyme activity which could be observed for fibers in a 0.1 M fructose assay mixture. Table 4 gives the corresponding values for membranes formed by casting the same solution on glass and air drying.

Results with Membranes

Secondary Acetate Membranes

Results for an early column which was packed with membranes which were cast on water and cut into small pieces are given in Figure 2. The activities observed in the column were low and the life of the membrane short.

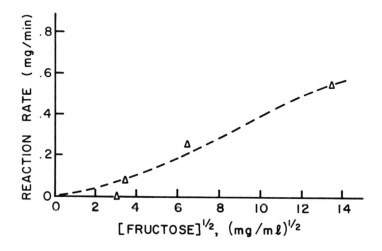

FIGURE 1. REACTION RATE vs. SUBSTRATE CONC.

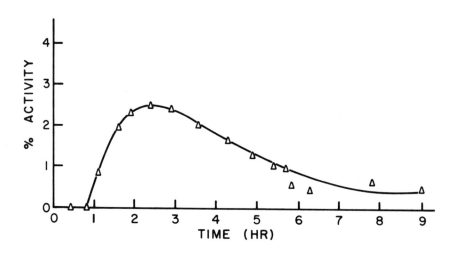

FIGURE 2. WORKING LIFE OF THE MEMBRANE

Table 3

SUMMARY OF THE EFFECTS OF SECONDARY
SOLVENT ADDITIONS FOR FIBERS

Solvent	Level	
	Low	High
Ethanol	2.3	--
Ethyl acetate	5.6	--
Toluene	7.6	--
Acetone	7.6	10.4
Ethylene dichloride	11.5	17.3

Table 4

SUMMARY OF THE EFFECTS OF SECONDARY
SOLVENT ADDITIONS FOR MEMBRANES

Solvent	Level	
	Low	High
Ethanol	45	--
Ethyl acetate	26	--
Toluene	20	--
Acetone	58	58
Ethylene dichloride	46	34

Table 5

RESULTS THAT SHOW CELLS ARE
LEAKING FROM THE MEMBRANES

Assay Number	Activity with Membranes Present	Activity after Membranes Removed
1	32.8	12.9
2	24.0	4.1
3	17.6	1.3
4	14.9	1.9
5	9.9	0.9

Table 6

COMPARISON OF VARIOUS METHODS OF
MAKING MEMBRANES

Type of Membrane	Activity with Membrane Present	Activity with Membrane Removed
Cast on water	26	13
Glass - air	21	3.4
Glass - wet	19	2.0

An experiment was performed to determine if cells were leaking from the membrane by some process. It was already determined that very little enzyme activity was capable of leaking from the cells. The membranes were assayed in a batch for one hour, then removed from the assay solution and any residual enzymatic activity determined. Table 5 gives the results which show that a substantial portion of the activity was leaking from the membranes.

Various ways of forming membranes were experimented with to see what effect the manner preparation had on the leakage of cells. Results are given in Table 6 for membranes formed with acetone as the solvent and formed in various ways. Glass cast membranes appear to be the best.

Various solvent systems were experimented with also. Table 7 gives results for membranes prepared from a 6:1 methylene dichloride: acetone solvent system, cast on glass and air dried. Note that this solvent system gives the highest initial activity and the lowest leakage for the secondary acetate.

Triacetate Membranes

Some of the triacetate membranes have been investigated for cell leakage. The results thus far are given in Table 8 which shows that the problem of cell leakage was also present and similar to that of the secondary acetate.

DISCUSSION

Our work with fibers shows that the rate of diffusion of substrate into the fiber is the cause of the limited activities observed. We have shown that permeability of the cellulose acetate can be improved by a proper choice of solvent systems; however, other approaches will probably also be required before the fiber achieves high observable activities at the whole cell levels which were used. Since cellulose acetates have been worked with by a large number of investigators in other areas, much of the technology needed to obtain desirable permeabilities already exists.

Our work with membranes resulted in some rather respectable activities. Although a problem of cells leaking from the membranes was encountered, considerable progress was made towards eliminating the problem for membranes made with secondary acetate, and further work can be undertaken for evaluating these membranes for industrial applications.

Our conclusion from the work is that the entrapping of whole cells in cellulose acetates appears to have the potential for being an inexpensive industrial method for immobilizing enzymes.

Table 7

RESULTS FROM ANOTHER SOLVENT SYSTEM

$6:1 - CH_2Cl_2:Acetone$

Activity with Membranes Present	Activity after Membranes Removed
46	5

Table 8

RESULTS SHOWING THE LEAKING OF CELLS
FROM CTA MEMBRANES

Type of Membrane	Activity with Membrane Present	Activity after Membrane Removed
Acetone - low	58	22
Ethanol - low	45	21

The authors wish to acknowledge the support of this work by the RANN Program of the National Science Foundation through Grant No. GI 34919. A special thanks goes to Mr. William H. McMullen III of Novo Enzyme Corporation for supplying some of the material used in this work.

REFERENCES

1. G.J.H. Melrose, Rev. Pure Appl. Chem., 21, 83 (1971).

2. F. Leuschner, British Patent 953,414 (1964).

3. W.R. Vieth, S.S. Wang, and R. Saini, Biotechnol. Bioeng., 15, 565 (1973).

4. D. Dinelli, Process Biochem., 7, 9 (1972).

GLUCOSE ISOMERASE:

A CASE STUDY OF ENZYME-CATALYZED PROCESS TECHNOLOGY*

Bruce K. Hamilton, Clark K. Colton, and Charles L. Cooney**

Department of Nutrition and Food Science
and
Department of Chemical Engineering
Massachusetts Institute of Technology
Cambridge, Massachusetts   02139

ABSTRACT

Process engineering considerations important in the exploitation
of enzyme-catalyzed reactions for large-scale production of desired
products are illustrated in the context of a case study of glucose
isomerase technology. The state of the art of glucose isomerase
processing as revealed by journal and patent literature is reviewed
and assessed. Among topics covered are enzyme production, immobilization and stabilization, kinetics, reactor design, and
product recovery. Some possible future processing objectives, such
as production of pure fructose, are discussed.

INTRODUCTION

The principal aim of biochemists working in enzymology is
elucidation of the patterns, and of the means of regulation, of
biochemical pathways and reaction mechanisms. In contrast, the
primary purpose of engineering is practical application of scientific knowledge, for example, the development of processes for
producing useful products. Since the objectives of engineers and
enzymologists differ so widely, it is not surprising that significant efforts may be required when an enzyme-catalyzed process is

*Publication number 2298 from the Department of Nutrition and
 Food Science, Massachusetts Institute of Technology.
**To whom correspondence should be addressed.

to be designed into a commercial reality, even if the enzymology
of the reactions involved is either well-known or relatively simple.
The purpose of this paper is to identify process engineering considerations important in the rational exploitation of enzyme-catalyzed reactions for large-scale production of desired products.
These considerations are illustrated in the context of a case study
of processes involving glucose isomerase, an enzyme in which there
is currently considerable interest in the food processing industry.
Glucose isomerase catalyzes the conversion of glucose to fructose
(see Figure 1).

Before becoming immersed in the details of glucose isomerase
processes, it is worthwhile to place the use of this enzyme into
perspective in relation to the entire field of enzyme technology.
The next section is intended to provide such an orientation.

Scope of Enzyme Technology: The Place of Glucose Isomerase

The wide range of enzyme technology applications is outlined
in Table I.   Enzymes are employed in areas which include large-
scale industrial processing in the food, paper, textile, pharmaceutical, fine chemical, and detergent industries, as well as in
the areas of chemical (especially clinical) analysis, treatment of
disease, and basic research in the physical, chemical, and
biological sciences.   Currently, there is interest in applying the
unique advantages of enzyme-catalyzed reactions to preparative-
scale stereospecific organic synthesis and drug discovery.   Within
this broad spectrum of activities, the application of glucose
isomerase falls into the category of large-scale industrial
processing.   Additional perspectives on enzyme technology have

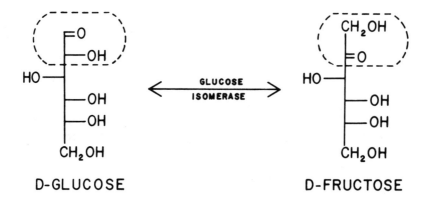

D-GLUCOSE   D-FRUCTOSE

Figure 1:   Isomerization of glucose to fructose.   Open chain
structural formulas are shown.

Table I

Categories of Enzyme Applications

Category	Selected References
General	Wolnak (1972), Beckhorn (1971), deBecze (1967), Sherwood (1966)

Large-Scale Processing

a. Food and Beverage Industries Example: Glucose Isomerase	Wieland (1972), Underkofler (1968), Reed (1966)
b. Paper and Textile Industries	Barfoed (1970)
c. Pharmaceutical and Fine Chemical Industries Examples: Penicillin amidase D-L amino acid resolution . .	Warburton, et al. (1973) Tosa, et al. (1969)
d. Detergent Additives	Wolnak (1972), Jensen (1972), Vogels (1971)
e. Waste Treatment	Slote (1970)
Chemical (especially clinical) Analysis	Cooney, et al. (1974), Inman and Hornby (1972), Guilbault (1970)
Enzyme Therapy	Dahlqvist, et al. (1973), Sizer (1972), Chang (1972)

Preparative-Scale Stereospecific
Organic Synthesis and Drug
Discovery

Examples: steroids	Mosbach and Larsson (1970)
amino acids	Smolarsky, et al. (1973)
antibiotics	Conover (1971)
Basic Research	Zaborsky (1973)

Table II

Glucose Isomerase in the Context of Basic Groups of Enzyme Reactions

Group	Selected References on New Applications
I. Degradations	
a. Soluble Substrates	
b. Insoluble Substrates	McLaren and Packer (1970); Archer, et al. (1973); Ollis, et al. (1973); Van Dyck, et al. (1974)
II. Simple Transformations	
a. Non-Cofactor Enzymes	
Example: Glucose Isomerase	
b. Cofactor (e.g., NADPH) Enzymes	
III. Syntheses (requiring, for example, ATP)	Marshall, et al. (1972); Berman and Murashige (1973); Hamilton, et al. (1974 a)

recently been given by Faith, et al. (1971), Carbonell and Kostin (1972), Wingard (1972a, 1972b), Zaborsky (1973), and also by papers in this volume.

Enzyme reactions can be classified into the three basic groups listed in Table II: (I) degradations, (II) simple transformations, and (III) syntheses. So far, most large-scale processing applications have involved only degradative reactions employing extracellular enzymes. In general, less work has been done on processing applications involving enzymes (often intracellular) which catalyze simple transformations, and research on applications involving enzymes (almost always intracellular) which catalyze synthesis reactions is only just beginning. Within this classification scheme, glucose isomerase is an intracellular enzyme which catalyzes a simple transformation, and it is exceptional in the context of the scheme just described in that its processing applications have been, and still are, extensively studied.

Given the preceding perspective, it is now appropriate to survey existing commercial glucose isomerase processes, and then to examine in more depth details associated with these processes. First, however, some comments on the information sources upon which this review is based.

INFORMATION SOURCES

The basis for this review is the published journal and patent literature. In focussing upon an enzyme in which there is current commercial interest, we recognize that proprietary technology and marketing information may be more advanced than that discussed here, and that we therefore run the risk of inviting gentle admonishments from our industrial colleagues who are involved in the serious business of using glucose isomerase in on-stream, large-scale processing systems. Recently, for example, B.J. Schnyder (1973), Assistant Manager of Research, Clinton Corn Processing Company, noted in an open letter that not only do several companies have the potential capability to make glucose isomerase at commercially acceptable costs, but also that: "The Clinton Corn Processing Company, a division of Standard Brands, Incorporated, has been commercially manufacturing and using immobilized glucose isomerase for several years." With this technology, Clinton and other companies are producing hundreds of millions of pounds of product each year at low cost. The most extensive published sources of information on U.S. commercial development of glucose isomerase processes are papers prepared by workers at Corn Processing Corporation (Kooi and Smith, 1972) and Clinton Corn Products (Harden, 1972, 1973; Newton and Wardrip, 1973), and patents cited below. It is precisely because so much work has already been done that production of high-fructose syrups and crystalline fructose is an excellent example with which to illustrate considerations important in the engineering of enzyme processes.

Summary Survey of Glucose Isomerase Processes

The first paper dealing with glucose isomerase was produced by the Americans Marshall and Kooi (1957); a U.S. patent based on this effort was issued in 1960 (Marshall, 1960). Japanese papers began with the work of Tsumura and Sato (1960), with other Japanese researchers, such as Yamanaka (1963a) and Takasaki (1966), publishing later. Only recently have American investigators again begun to publish work on glucose isomerase (e.g., Strandbert and Smiley, 1971, 1972; Vieth, et al., 1973; Wang and Vieth, 1973; Havewala and Pitcher, 1974; publications by workers at Corn Processing Corporation, and Clinton Corn Products cited above). Continuing interest in the study of glucose isomerase in this country is further exemplified by several papers within this volume discussing its immobilization (Olson and Stanley, 1974); Bernath and Vieth, 1974; Kolarik, et al., 1974). A chronology of many important isomerase patents appears in Table VI and illustrates the increasing interest in this enzyme; this applied work will be reviewed in the following sections.

American industry today has purchased isomerase technology from the Japanese (Chem. Eng. News, 1973), as well as developed considerable technology itself. Interestingly, American investigators (Marshall and Kooi, 1957) reported discovery of glucose isomerase activity several years before Japanese workers began publishing in the area (Tsumura and Sato, 1960). This circumstance may reflect some economic advantages in licensing foreign technology as compared to developing domestic technology from scratch. Perhaps in the area of strain acquisition and genetic development, licensing is very much cheaper. A published isomerase process flow sheet is shown in Figure 2, and typical product specifications are given in Table III. In Clinton's process, corn starch is first converted to a refined dextrose liquor (top half of Figure 2) using an acid/enzyme process. The dextrose is then isomerized, refined, and concentrated to yield the final product. Saccarification and feed preparation are operated in batch mode, liquefaction and the two evaporation steps are completely continuous, and remaining sections are all semi-continuous Harden, 1972).

Components of Enzyme Process Development

Except for plant design aspects, the basic components of enzyme-catalyzed process development are listed in Table IV. Each of these components is illustrated and discussed in the context of glucose isomerase technology in the remainder of this review.

1. MARKET ANALYSIS AND TECHNOLOGY ASSESSMENT

A market analysis showing acceptable profit potential, coupled with a favorable review of competing production processes, is undoubtedly what any business manager would like to start with before permitting enzyme process development to begin (Lantos, 1973). From published information, the market potential for products produced using glucose isomerase is relatively clear. Basically, these products fall into two categories: glucose/fructose syrups, and pure fructose.

Since the single-pass equilibrium conversion of glucose to fructose is about 50 percent (see a later section for more detail), production of pure fructose from glucose using glucose isomerase involves a product separation step . However, production of fructose/glucose syrups, sweeter than glucose syrups, involves no product separation step and is therefore attractive. Furthermore, according to a recent study (Wolnak, 1972) fructose/glucose

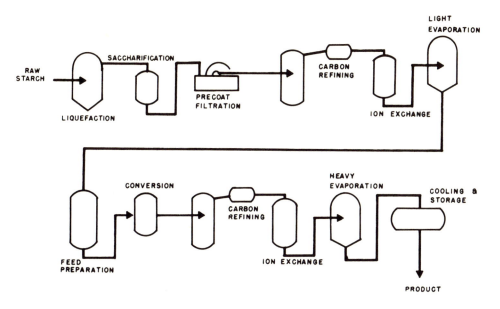

Figure 2:   Flowsheet of industrial process (Clinton Corn Processing
Company)for production of high-fructose corn syrups
using glucose isomerase (source: Harden, 1972).

Table III

Typical Analysis of Isomerose 100
(Clinton Corn Processing Company)

Solids	71%
Moisture	29%
pH	4.3
Pounds per Gallon @ 100°F	11.1
Pounds Solids per Gallon @ 100°F	7.9
Color (CIRF × 100)	0.4

Dry Basis Composition:

Ash (Sulfated)	0.03%
Carbohydrate Components	
Dextrose	50%
Fructose (levulose)	42%
Other saccharides	8%
Relative Sweetness*	100

*Compared to sucrose in a 15% solids solution

Table IV

Basic Components of Process Development
for Enzyme-Catalyzed Reactions

1. Market Analysis and Technology Assessment

2. Comparison with Competing Processes

3. Technical Considerations

    a. Basic Enzymology

    b. Enzyme Production (Microbial Enzymes)

        1. Microorganism Acquisition and Strain Development
            (Genetic Manipulation)
        2. Enzyme Fermentation and Environmental Manipulation
        3. Cell Disruption or Enzyme Fixing (Intracellular
            Enzymes)
        4. Enzyme Purification and Recovery
        5. Enzyme Storage, Handling, and Toxicity

    c. Enzyme-Catalyzed Reaction

        1. Enzyme Stabilization
        2. Microbial Contamination
        3. Thermodynamics
        4. Solution Kinetics
        5. Immobilization
        6. Immobilized Enzyme Kinetics
        7. Product Lability
        8. Reactor Design

    d. Enzyme Reuse or Recycle

        1. Immobilization
        2. Ultrafiltration

    e. Product Purification

syrups produced with glucose isomerase "....have about the same
degree of sweetness as invert sugar, but since they are cheaper
(Table V), will probably replace invert sugar in most food
applications in the future." The present U.S. market for invert
sugar is in excess of two billion pounds per year (Wolnak, 1972).
Details on some of the food applications of high-fructose syrups

Table V

Recent Selling Prices of Some Sweet Syrups and Crystalline Sugars

Sweet Syrups	$/lb., dry basis
Invert sugar (40% fructose, 40% glucose, 20% sucrose)	0.12*
Glucose/fructose syrup (Isomerase^R 100)	0.11*

Crystalline Sugars	
Fructose	1.20**
Glucose	0.135**
Sucrose	0.127**

---

 * Wolnak (1972).
** Chemical Marketing Reporter, October 29, 1973.

are given in the paper by Newton and Wardrip (1973), and a more general perspective on use of sweetners in foods has been presented recently by Nicol (1971).

Non-food applications of fructose include use in parenteral solutions for intraveneous infusion and use as a chelating agent (Ward, 1967). Recently, a patent (Gordon, 1971) on application of fructose as a de-intoxicant was issued, but there is controversy about efficacy (Chem. Eng. News, 1971a, 1971b).

Recent selling prices of food grade fructose are very much higher than those of either glucose or sucrose (see Table V). Since fructose can be considered to be a natural low calorie sweetner (Nicol, 1971), a significant domestic market might develop if low cost industrial production of fructose became feasible. At least until the late sixties, the only producer of pure fructose in the United States was Dawe's Laboratories, Inc., (Ward, 1967). At that time, Dawe's process was based on a method disclosed by Holstein and Holsing (1962) for the separation of fructose from invert sugar solutions using glucose oxidase; this method is discussed in a later section.

Technology assessment methodology has been applied by Rubin (1971) to forecast possible indirect societal impacts of expanding glucose isomerase markets. Rubin concludes that economic, demographic, and even political (both domestic and international)

effects could be great if fructose/glucose syrups produced from
corn or potato starch displaced a significant portion of the cane
and beet sugar market.

2. COMPETING PROCESSES

Extensive work on alkaline isomerization of glucose to fructose
has been performed (e.g., Fetzer and Evans, 1949; Langlois and
Larson, 1956; Mendincino, 1960; Haack, et al. 1964; Kainuma, et al.,
1964, 1966a, 1966b, 1967a, 1967b, 1968a, 1968b, 1968c; Scallett,
et al., 1967, 1968, 1972; Barker, et al., 1973a, 1973b). Shift of
equilibrium yield in favor of fructose has been effected using
borate salts (Mendincino, 1960), borate resins (Barker, et al.,
1973b, and aluminate salts (Haack, et al., 1964). Apparently, the
alkaline route has not been successfully commercialized because of
color, flavor, and composition problems (Lloyd, et al., 1972;
Newton and Wardrip, 1973).

The German firm C.F. Boehringer has developed a process for
the production of fructose from sucrose (Boehringer & Sons, 1967,
1968). In the Boehringer process, sucrose is hydrolyzed to glucose
and fructose in a column packed with a sulfonated polystyrene resin
which is partially in the calcium form. The glucose and fructose
are eluted separately after chromatographic separation on the same
column, and fructose may then be crystallized. To our knowledge,
this technology has not yet been introduced for commercial
production of fructose within the United States, although evidently
it is used for fairly sizable fructose production in Europe.

3. TECHNICAL CONSIDERATIONS

Many of the technical considerations listed in Table IV are
interactive with each other. For example, isomerase stabilization
might be a by-product of enzyme immobilization on an artificial
support or of enzyme fixation to cells. The approach to kinetic
studies and reactor design would likely be different depending upon
whether isomerase is used free in solution and is recycled by
ultrafiltration, is immobilized, or is fixed in cells which are
used free in suspension or are themselves immobilized. Interplay
of such factors should therefore be kept in mind as the following
discussion procedes.

Basic Enzymology

Two types of glucose isomerase activities have been established.
As summarized by Noltmann (1972): "Numerous microorganisms were

Table VI

Microorganisms Used as Sources for Glucose Isomerase in Some Patent Literature Examples

Microorganisms	Authors	Organization	Patent
Pseudomonas hydrophila	Marshall (1960)	Corn Products Company	U.S. 2,950,228
Streptomyces flavovirens " achromogenes " echinatus " albus " bobiliae	Takasaki, et al. (1966)		Japanese 7428(66) " 7430(66) " 7431(66) British 1,103,394
Streptomyces phaeochromogenes " fradiae	Sato and Tsumura (1966)		Japanese 17,640 (66)
Streptomyces phaeochromogenes " fradiae		Food Res. Inst., Tokyo	Japanese 28,473 (69)
Streptomyces wedmorensis (ATCC 21230) " flavovirens (ATCC 3320) " achromogenes (ATCC 12767) " echinatus (ATCC 21933) " albus (ATCC 21132)	Takasaki and Tanabe (1971)	Agency of Industrial Science and Technology, Tokyo	U.S. 3,616,221
Streptomyces species (ATCC 21173)	Cotter, et al. (1971)	Standard Brands, Inc.	U.S. 3,623,953
Arthrobacter nov. sp. (NRRL B-3728) " nov. sp. (NRRL B-3726) " nov. sp. (NRRL B-3727) " nov. sp. (NRRL B-3724) " nov. sp. (NRRL B-3725)	Lee, et al. (1972)	R.J. Reynolds Tobacco Company	U.S. 3,645,848
Streptomyces sp. (ATCC 21175) " sp. (ATCC 21175)	Dworschack and Lamm (1972) Lloyd, et al. (1972)	Standard Brands, Inc. Standard Brands, Inc.	U.S. 3,666,628 U.S. 3,694,314
Streptomyces albus (ATCC 21132)	Takasaki and Kambayashi (1972)	Bureau of Industrial Technology, Japan	Japanese 19,086 (72)
Streptomyces olivaceus (NRRL 3916) " olivaceus	Brownwell and Streets (1972) Zienty (1972)	Miles Labs., Inc. " " "	German 2,219,713 German 2,223,340
Streptomyces olivochromogenes (ATCC 21,114)	Heady and Jacaway (1972)	CPC International, Inc.	German 2,223,864
Streptomyces phaeochromogenes Lactobacillus brevis	Sipos (1973)	Baxter Labs., Inc.	U.S. 3,708,397
Nocardia asteroides Nocardia dassonvillei Micromonospora coerula Micromonospora rosea Microellobospora flavea	Horwath and Cole (1973)	Standard Brands, Inc.	German 2,247,922
Streptomyces sp. (ATCC 21175) " " "	Dworschack, et al. (1973a) " " " (1973b)	Standard Brands, Inc. " " "	German 2,251,855 U.S. 3,736,232
Streptomyces albus	Takasaki (1973)	Inst. Ferm., Osaka	Japanese 49,981 (73)

screened for D-glucose isomerizing activity, but a genuine, specific
D-glucose isomerase apparently does not exist. Instead, most of
the enzymes in this category turned out to be either D-xylose
isomerase or glucose-6-phosphate isomerase." The latter evidently
requires arsenate to act on non-phosphorylated glucose. Since the
activity discovered by Marshall and Kooi (1957), and also Tsumura
and Sato (1960), apparently required arsenate, the enzyme responsible
for isomerization in their cases probably was glucose-6-phosphate
isomerase. (Also note that the microorganism used as enzyme source
by Marshall and Kooi was Pseudomonas hydrophila, NRC 491 and 492,
and that used by Tsumura and Sato was Aerobacter cloacae. NRC 491
and 492 have been reclassified as Aerobacter cloacae [Clement and
Gibbons, [1960]).The isomerases studied by Yamanaka (1963a),
Takasaki (1966), and Danno (1970a),however, did not require
arsenate for activity and were actually xylose isomerases. Because
the primary market for glucose/fructose syrups is in food applications, the requirement for arsenate makes glucose-6-phosphate
isomerase less attractive than xylose isomerase. It is not
surprising, then, that most of the development work after 1960
focused on the latter enzyme, as is apparent upon examination of
the contents of the patents listed in Table VI. Further discussion
of the basic enzymology of xylose and glucose-6-phosphate isomerases
is given by Noltmann (1972).

Enzyme Production

Microorganism acquisition and strain development (genetic
manipulation). All reported sources of glucose isomerase are
microbial. Yamanaka (1968) and Newton and Wardrip (1973) have
provided compilations of references on D-glucose isomerase activities of many microorganisms. Microorganisms used in some patent
examples are listed in Table VI. Increasing interest in glucose
isomerase is apparent from the chronological frequency of patents.
Also evident is the predominant use of Streptomyces species as
enzyme source, most likely a consequence of the fact that enzyme
from these species is the xylose isomerase.

Once an organism is isolated for use as enzyme source, the
objective becomes maximization of enzyme content in the isolate
when it is cultured in a production fermentor. Genetic manipulations
to increase in vivo levels of enzymes have been discussed in general
by Demain (1971a, 1971b, 1972). In the case of glucose isomerase,
Bengtson and Lamm (1972) employed a random mutation/screening
program to obtain Streptomyces cultures capable of improved isomerase production. Newton and Wardrip (1973) describe the work of
Bengtson and Lamm as follows: "... (Toxic) agents were added to
the viable microorganisms so that 90-95% of the microorganisms were
inactivated. The term 'microorganism' in the patent includes both

spores and vegetative cells. Actively growing Streptomyces
cultures were treated with such chemicals as ethyleneimine, hydrogen
peroxide, 8-ethoxycaffeine, as well as by radiation from radio-
active isotopes and ultraviolet light. The surviving cells were
isolated, cultured and checked for their ability to produce
isomerase. They claim that an increase of 30-50% in the yield of
glucose isomerase can be obtained from colonies isolated from
treated cultures."

Screening for new organisms and over-producing mutants is
labor intensive and consequently expensive. Tsumura, et al. (1967)
reported that they developed a rapid assay suitable for screening
microbial colonies for improved glucose isomerase production.
Their approach was to add a loop of cells to a piece of filter
paper which was saturated with a 1% buffered glucose solution and
then placed onto a second piece of filter paper. After an hour of
incubation, the second piece of filter paper was assayed for fructose. In this manner, many cultures could be screened rapidly.
It is conceivable that such a procedure might be automated.

Enzyme fermentation and environmental manipulation. Environmental manipulations to increase in vivo levels of enzymes have
also been discussed in general by Demain (1971a, 1971b, 1972), and
the influence of nutritional environment on enzyme production has
been reviewed by Dean (1972). In the case of glucose isomerase
fermentations, two interesting categories of environmental
manipulations which have been investigated are medium composition
and morphology.

In the category of medium composition, various constituents
have been found to stimulate production of glucose isomerase.
First, it should be recalled that glucose isomerase acts not only
on glucose but also (and preferentially) on xylose (Yamanaka, 1963b;
Takasaki et al., 1969a). In fact, most reported microorganisms
require in their growth medium a source of xylose for induction of
glucose isomerase activity. In some cases, expensive xylose can
be supplied indirectly in the cheaper form of xylan or xylan-
containing materials such as straw, wheat bran, corn cobs, corn
husks, or pulp waste (Takasaki, et al., 1969b and 1971). The
utilization of these cheap raw materials greatly reduces the cost
of the fermentation broth. Lee, et al. (1972) reported that three
strains of Arthrobacter (NRRL B-3726, NRRL B-3727, NRRL B-3728)
produce high levels of isomerase in the absence of added xylose or
xylan when grown on a medium containing meat protein, yeast
extract, glucose, and inorganic salts. However, since xylan-
containing materials are readily and cheaply available, an expensive meat protein/yeast extract broth would certainly seem to
offer no advantage.

Takasaki (1966, 1973) found that cellular levels of glucose isomerase in bacteria (e.g., Streptomyces albus) and yeast were enhanced by Co^{2+}. Yoshimura, et al. (1966) reported that several metallic salts ($CoCl_2$, $MnSO_4$, $ZnSO_4$) enhanced isomerase production in Bacillus coagulans, but some ($CoCl_2$, $ZnSO_4$) depressed growth. As discussed below, several inorganic ions are activators of isomerase once the enzyme is recovered.

Dworschack, et al. (1973b) reported that sorbitol enhanced production of glucose isomerase in Streptomyces. Along these lines, Danno (1970b) found that sorbitol is a competitive inhibitor of isomerase recovered from Bacillus coagulans, and Takasaki, et al. (1969a) reported that it is also a competitive inhibitor of isomerase recovered from Streptomyces albus. Danno (1970a) found that threonine and glycine stimulated isomerase production in Bacillus coagulans, and Heady and Jacaway (1972) reported that glycine and NH_4NO_3 stimulated isomerase production is Streptomyces olivochromogenes ATCC 21114. Takasaki and Kambayashi (1972) reported raising isomerase yields in S. albus ATCC 21132 cultures by first incubating for 30 hours and then making sequential additions of xylan at one-hour intervals for eight hours. In all these cases, it is clear that the approach and intent is to optimize the growth environment with regard to maximizing enzyme productivity (units of enzyme activity/fermentor/time) per unit of operating cost, which includes expenses arising from growth medium and fermentation time.

In the category of morphological manipulation, Dworschack and Lamm (1972) reported an innovation which they claimed increased isomerase levels by factors of three to four. They pointed out that microorganisms of the Streptomyces genus, often used to produce glucose isomerase, have a tendency to grow in the form of compact spherical pellets. When small amounts of agar, carboxymethyl cellulose, or diatomaceous earth were dispersed in the growth medium, mycelia produced were filamentous, and isomerase activity levels (units/ml broth) were increased. While the physiological reasons for this increase in production are not clear, it is possible that growth in the diffuse mycelial form permits better exchange of oxygen and other nutrients with the cell than does pellet growth.

Cell disruption or enzyme fixing. Glucose isomerase is an intracellular enzyme. Several methods for its release have been reported. Sonication has been used by many laboratory workers (e.g., Tsumura and Sato, 1961; Takasaki, 1966; Strandberg and Smiley, 1971), but this method has not yet proved feasible for large-scale preparations. Takasaki, et al. (1969a) reported that glucose isomerase was easily liberated from their Streptomyces strain by autolyzing at 40°C around pH 6.5, and that when a

cationic surface active agent, such as cetyl pyridinium chloride,
was added, the rate of liberation was markedly increased.
According to Sipos (1973), addition of 4% toluene to S.
phaeochromogenes in fermentation broth initiates isomerase
release which can be augmented by lysozyme/EDTA treatment.

It is not necessary, however, for isomerase to be
completely released. For example, according to Takasaki
and Tanabe (1971), disintegrated S. wedmorensis cells
prepared by passing an aqueous cell suspension through a
high-pressure homogenizer can be used directly as a catalyst.
Most interestingly, Takasaki, et al. (1969b) found that
when their Streptomyces species was heated at temperatures
above 60°C for about 10 minutes, isomerase was fixed or stabilized within and/or on the cells. This heat-fixed enzyme
was not liberated even if incubated under conditions favorable
for autolysis (40°C, pH 6.5). A modified heat-fixing procedure
is discussed in a patent by Lloyd et al. (1972) covering work
with isomerase from a Streptomyces species. Optimum results
with this fixation or stabilization treatment were obtained
if whole cells (before or after being separated from fermentor
broth) were heated at temperatures ranging from about 70° to
80°C, at a pH ranging from about 7 to 8, for 10 to 20 minutes.
Lloyd, et al. (1972) hypothesized that the basis for isomerase
stabilization was that cellular enzymes responsible for
autolysis of the cells (which results in leakage of isomerase
out of the cell) are inactivated by heat treatment. They
also pointed out that heat stable isomerase (apparently
D-xylose isomerase) was preferred over the heat labile (glucose-
6-phosphate isomerase) enzyme which requires arsenate or
flouride ion for activity with glucose.

Enzyme purification and recovery. Three independent
groups have purified glucose isomerase to homogeneity as
determined by sedimentation and/or electrophoresis. The steps
of their purification schemes are shown in Table VII. Glucose
isomerase from Streptomyces (Takasaki, et al., 1969a, 1969b)
has been purified to homogeneity as determined both by sedimentation and electrophoresis, and it is interesting to note that
an overall purification factor of about 10 resulted in an
apparently pure enzyme. This suggests that about 10% of
the cell protein released by autolysis was composed of isomerase,
an amount which seems relatively high. The implication of this
for enzyme production is that such sources of isomerase already
have a relatively good specific activity and as a consequence
may require little or no purification. In fact, glucose
isomerase immobilized on DEAE-cellulose (Sipos, 1973) was not
subjected to any purification operations except removal of

Table VII

Schemes for Purification of Glucose Isomerase

STEP	Yamanaka (1968)	purification factor	Tasakasi et al. (1969a, 1969b)	purification factor	Danno (1970a)	purification factor
1	Extraction by grinding with alumina	1.0	Extraction by autolysis at 40°C, pH 6.5	1.0	Extraction with toluene and lysozyme	1.0
2	Mn^{2+} treatment	1.9	Acetone	1.7	Mn^{2+} treatment	1.5
3	Ammonium sulfate I	3.0	DEAE-cellulose	6.8	Ammonium sulfate	11
4	Heat treatment	5.2	DEAE-sephadex	9.9	DEAE-Sephadex I	43
5	Acetone	7.1	1st crystals	9.0	DEAE-Sephadex II	43
6	DEAE-Sephadex	9.5	2nd crystals	10.0	Ammonium sulfate	43
7	Ammonium sulfate II	10.0	Sephadex G-200	10.0	Acetone crystals	43
8	1st crystals	10.9				
9	2nd crystals	11.5				
Homogeneity Tests	Sedimentation		Sedimentation, electrophoresis		Sedimentation, electrophoresis	
Activity Recovery	21.7%		<3.4%		36.6%	
Microorganism	Lactobacillus brevis		Streptomyces albus		Bacillus coagulans	

*purification factor for step n = specific activity nth step / specific activity 1st step

fermentor broth solids by screening and cell-debris by filtering
with filter aid. When isomerase is used in the cell-fixed form,
again obviously no purification is employed. From a practical
viewpoint, it is desirable to employ an enzyme preparation
produced using as few processing and purification steps as
possible in order to minimize cost and loss of total isomerase
activity.

          Enzyme storage, handling, and toxicity. The question of
enzyme stability during operation is treated later and the point
of concern here is storage prior to use or immobilization. An
enzyme with a storage half-life measured in days is difficult to
work with; however, if storage half-life is measured in months,
then it is possible to prepare and store large batches of enzyme
for use as required. Lee, et al. (1972) noted that centrifuged
isomerase-containing cells could be stored frozen at -5°C for
extended periods of time. Strandberg and Smiley (1971) reported
that lyophilized isomerase retained more than 70% of its activity
after 1.5 years of storage at 4°C. It therefore appears that
isomerase from Streptomyces is quite stable during storage under
proper conditions.

          Another aspect of enzyme production, handling, and storage
is health and safety of personnel. During development of detergent
enzymes a number of problems arose in connection with plant
personnel handling large quantities of proteases. For this reason,
biosafety procedures such as those outlined in general by Dunnill
(1974) should govern manufacturing operations.

          Finally, there is the question of safety with regard to the
final products of isomerase processes. New processes for the
production of foods or food additives require FDA approval and
proof of lack of any toxic or otherwise harmful effects. Kooi
and Smith (1972) have reported feeding studies with Streptomyces
olivochromogenes which demonstrate that this source of isomerase
is not toxic and that isomerase prepared from this organism does
not elicit a deleterious response in animals. Undoubtedly similar
tests have been run on other strains and enzyme preparations
obtained from them.

                         Enzyme-Catalyzed Reaction

          Enzyme stabilization. Stabilization of enzyme activity under
reaction conditions is an extremely important economic factor.
Commonly, attempts are made to stabilize enzymes by immobilization,
or, in the particular case of glucose isomerase, by fixing enzyme
to cells (which may in turn be immobilized). Caution should be
exercised when evaluating the stability of such preparations because

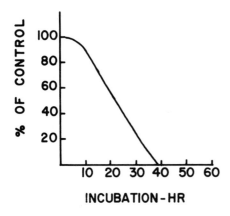

Figure 3: Stability of
glucose isomerase in
solution, at 70°C
(Sipos, 1973).

it is possible for diffusional resistances or microenvironmental
effects to cause an apparent increase in immobilized enzyme stability over native enzyme stability when actually there may be no
real stability increase at all. The contributions of internal
pore or gell diffusion and microenvironmental effects to this
phenomenon of "disguised instability" are discussed by Ollis
(1972). External, or "film" diffusion, can also contribute to
this phenomenon. Suffice it to say at this point that the objective is to stabilize enzyme activity in the simplest and cheapest
manner possible.

Figure 3, taken from Sipos (1973) shows that glucose isomerase
recovered from S. phaeochromogenes was relatively unstable when
free in solution at 70°C. Sipos also reported preparation of
immobilized glucose isomerase by adsorption onto DEAE-cellulose,
and the stability of this immobilized preparation at 70°C is shown

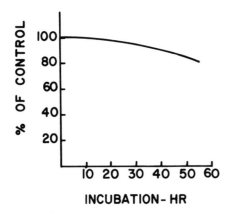

Figure 4: Stability of
glucose isomerase im-
mobilized by adsorption
on DEAE-cellulose, at
70°C (Sipos, 1973).

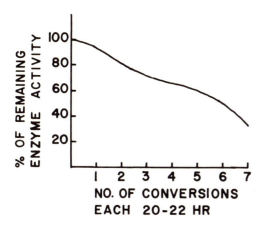

Figure 5: Stability of DEAE-cellulose-glucose isomerase after repeated reuse, at 70°C (Sipos, 1973)

in Figures 4 and 5. Assuming diffusional and microenvironmental influences on apparent stability were insignificant, stability of immobilized isomerase was much improved over that of isomerase free in solution. Figure 6 shows data obtained by Strandberg and Smiley (1972) using isomerase (again recovered from S. phaeochromogenes) bound to an aminoarylsilane derivative of porous glass beads packed in a column. These authors claimed that the erratic behavior of Column 1 (Figure 6) was primarily due to channeling of substrate solution through the bed, and that frequent agitation of the bed, as was done with Column 2 (Figure 6), reduced this erratic behavior. They also stated that the apparent loss of activity experienced with both columns might be due to dissolution of the exposed surface of the porous glass, where, presumably, the enzyme was covalently linked, and that Corning Glass Works reported a coated porous glass that is much more resistant to dissolution than uncoated glass. Very recently, two Corning workers (Havewala and Pitcher, 1974) immobilized isomerase on a

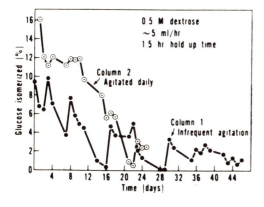

Figure 6: Stability of glucose isomerase bound to aminoaryl-silane derivative of porous glass (Strandberg and Smiley, 1972).

silanized derivative of zirconia-coated porous glass and reported
the stability data shown in Figure 7. If differences in assay
conditions (e.g., 60°C vs. 70°C) can be ignored, it seems
apparent that the stability of the isomerase immobilized on the
coated glass was better than that of isomerase immobilized on
DEAE-cellulose (Sipos, 1973).

When glucose isomerase is fixed to cells as described above,
activity evidently is also stabilized. Data presented by
Takasaki, et al. (1969b), obtained by cycling cell-fixed enzyme
through seven batch glucose isomerizations at 70°C, are shown in
Figure 8. The fall of activity with batch number in this case
is comparable to that observed by Sipos (1973) using isomerase

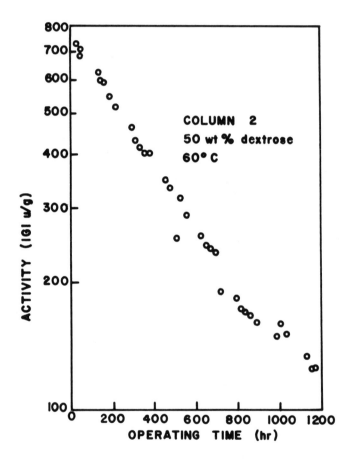

Figure 7: Stability of glucose isomerase bound to silanized derivative of zirconia coated porous glass (Havawela and Pitcher, 1974).

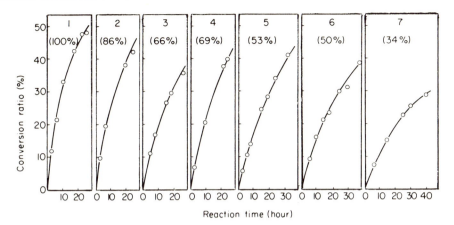

Figure 8: Use of recycled cell-fixed isomerase. Remaining
enzyme activity for each reaction (figure in parentheses) was
calculated from initial rate of each reaction. Reaction temperature was 70°C (Takasaki, et al., 1969b).

immobilized on DEAE-cellulose (Figure 5). When evaluating these
two methods of stabilization (cell-free immobilized isomerase vs.
cell-fixed isomerase) it should be kept in mind that Takasaki, et
al. claimed that 80-90% of the total glucose isomerase present in
a cell culture could be cell-fixed, while the corresponding figure
for immobilization procedures typically appears to be lower than
40% (Sipos, 1973; Havewala and Pitcher, 1974; Strandberg and
Smiley, 1971, 1972), although Giovenco, et al. (1973) claimed
that no isomerase activity was lost upon immobilization in spun
cellulose triacetate fibers.

     In an extension of the approach involving isomerase fixed
to cells, Vieth, et al. (1973) heat-treated S. phaeochromogenes
cells containing active glucose isomerase at 80°C for 1 1/2 hours,
and then immobilized these cells on collagen. The collagen-
immobilized cells were packed in a column which was used
continuously for 40 days at 70°C for glucose isomerization.
Results are shown in Figure 9. Initial conversion was 40%
(significantly below equilibrium conversion) and slowly decreased
over the 40 day period to about 30%. Apparent stabilization of
isomerase activity was excellent, but the extent of diffusional
and microenvironmental influences on apparent stability was not
made clear. In somewhat similar work, Kolarik, et al. (1974)
report in this volume on entrapment of glucose isomerase cells
in cellulose acetate membranes.

     Except for inorganic ions like Mn^{2+}, Co^{2+}, and Ni^{2+}
(Takasaki, 1966; Danno, 1970b), no chemical stabilizers of isomerase

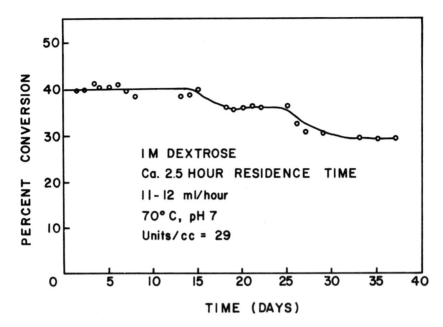

Figure 9: Performance of a column containing heat-treated S̲.
phaeochromogenes immobilized on collagen (Vieth, et̲ al̲., 1973).

have been reported, even though, in general, chemical stabilization
is not uncommon (Wiseman, 1973). For example, in the case of
acetate kinase, Gardner, et̲ al̲. (1974) have increased the half-life
in solution at 34°C from about one day to about 70 days by sequential addition of dithiothreitol. Since acetate kinase contains
sulfhydryl groups, this stabilization probably results from
maintenance of a reducing environment, and the same methodology
would not be expected to be effective for an enzyme insensitive
to oxidation. It is possible, however, that other chemical
stabilizers (e.g., glycerol, ethylene glycol) might be effective.
When this option is considered, it should be recognized that some
chemical stabilizers may be unacceptable for food processing.
Other approaches to stabilization include the utilization of
thermostable enzymes from thermophiles (Doig, 1974) and intramolecular crosslinking of the enzyme (Zaborsky, 1974).

    Microbial contamination. Microbial contamination of enzyme
processes can be disastrous, especially when dealing with food
materials. In fact, many food processes cannot be run continuously
because of this problem. By running at elevated temperatures,

Table VIII

Thermodynamic Data on Isomerization of Glucose to Fructose

A. Equilibrium Constants (in terms on concentrations,
   Keq = [fructose]/[glucose])

Temperature (°C)	Keq	
	Takasaki (1967)*	Havewala & Pitcher (1974)**
25	0.74	----
40	0.92	----
60	1.15	1.02
70	1.30	1.06
79.5	----	1.10

B. Heat of Reaction (Calculated using Van't Hoff equation)

Takasaki (1967)* . . . . . . +2.22 kcal/mole

Havewala & Pitcher (1974)**. +0.91 kcal/mole

---

* 0.00055 M sugar solutions, 0.045 M phosphate buffer (pH 7.0),
  0.0091 M $MgSO_4$

** 2.0 M sugar solutions, 0.2 M sodium maleate buffer (pH 6.85),
   0.02 M $MgSO_4$, 0.001M $CaCl_2$

such as about 70°C which is used with glucose isomerase, microbial contamination can be minimized if not eliminated. This approach to overcoming contamination is made possible by the increased stability of immobilized or cell-fixed isomerase. Another general approach to the problem is to use extremes of pH or both temperature and pH (Archer, et al., 1973).

Thermodynamics. Takasaki (1967) ran equilibrium isomerizations in the forward and reverse directions at different temperatures and reported the equilibrium constant data listed in Table VIII. When plotted according to the Van't Hoff equation (Weber and Meissner, 1967), these data yielded a straight line with a slope corresponding to an endothermic heat of reaction of 2.22 kcal/mole. Takasaki's equilibrium constant data were determined using dilute sugar solutions (about 0.00055 M).

Havewala and Pitcher (1974) also reported equilibrium composition data at different temperatures from which the equilibrium constants listed in Table VIII were calculated. Their experiments were carried out using relatively concentrated sugar solutions (2.0M). The calculated heat of reaction in this case was 0.91 kcal/mole.

Since the equilibrium constants listed in Table VIII are all written in terms of equilibrium concentrations rather than thermodynamic activities, the apparent differences in the two sets of thermodynamic data might arise from variation of activity coefficients in dilute and concentrated sugar solutions, or from chelation effects (see below).

Takasaki(1971) has demonstrated that equilibrium yield can be shifted in favor of fructose by incorporating chelating, or "trapping" agents into enzymatic isomerization mixtures. Takasaki found that 88% of glucose present at an initial concentration of 1.0 M was converted to fructose when enzymatic isomerization was carried out at around pH 7.5 in the presence of 0.3 M sodium tetraborate. However, if trapping agents such as this are used commercially, they must be removed from the final product if it is used in food applications.

Solution kinetics. Takasaki (1967) performed initial rate studies in dilute sugar solutions (<5%) at two temperatures in both forward and reverse directions using partially purified glucose isomerase extracted from Streptomyces albus. In all cases he obtained linear Lineweaver-Burk plots, and the Michaelis constants he determined from these plots are listed in Table IX. Michaelis constants determined by Yamanaka (1968) and Danno (1970b) for isomerase obtained from different microorganisms are also listed in Table IX.

These initial rate data support the hypothesis that in dilute sugar solutions glucose isomerization follows simple reversible Michaelis-Menten kinetics (i.e., no substrate or product inhibition or activation effects other than those associated directly with reverse reaction). If this is actually the case, then the reaction velocity expression is given by (Haldane, 1930; Peller and Alberty, 1959):

$$v = \frac{\dfrac{V_s}{K_s} s - \dfrac{V_p}{K_p} p}{1 + \dfrac{s}{K_s} + \dfrac{p}{K_p}} \qquad (1)$$

Table IX

Michaelis Constants for Forward and Reverse Glucose
Isomerase Reactions at Various Temperatures*

A. Takasaki (1967)

   microorganism: Streptomyces albus.

   conditions: pH 7.0 (45 mM phosphate buffer), 9 mM MgSO$_4$.

Temp (°C)	Km(forward)	Km(reverse)
25	0.062M	0.16M
60	0.14M	0.23M

B. Yamanaka (1968)

   microorganism: Lactobacillus brevis.

   conditions: pH 6.0 (30 mM maleate buffer), 0.5 mM MnCl$_2$,
                   0.5 mM CoCl$_2$.

   Km(forward) = 0.92M at 50°C

C. Danno (1970b)

   microorganism: Bacillus coagulans.

   conditions: pH 7.0 (7 mM barbital buffer), 10 mM CoCl$_2$.

   Km(forward) = 0.09M at 40°C

---

*Obtained using isomerase recovered from the indicated micro-
organisms and purified to homogeneity.

where s and p are substrate and product concentrations, K_s and K_p
are Michaelis constants for substrate and product, and V_s and V_p
are maximal forward and reverse reaction velocities. The magnitudes
of V_s and V_p depend on isomerase purity, but if Equation (1) is
followed, their relative values can be determined from the Haldane
relation when values of the Michaelis constants and the thermodynamic equilibrium constant are known. The Haldane relation can
be obtained by setting the velocity in Equation (1) to zero (the
equilibrium condition), which gives (Haldane, 1930):

$$K_{eq} = \left(\frac{V_s}{V_p}\right)\left(\frac{K_p}{K_s}\right) \qquad (2)$$

If the reaction velocity expression given by Equation (1) holds, and if there is no inactivation of enzyme with time, then an expression for the time course of an ideal (diffusion-free) batch isomerization can be obtained by intergrating Equation (1) to yield (Alberty, 1959):

$$
\left[\frac{V_s}{K_s} + \frac{V_p}{K_p}\right] t = \left[\frac{1}{K_s} - \frac{1}{K_p}\right] p - \left[1 + \frac{s_o}{K_p} + \frac{\left(\frac{1}{K_s} - \frac{1}{K_p}\right)}{\left(\frac{V_s}{K_s} + \frac{V_p}{K_p}\right)}\left(\frac{V_p}{K_p}\right) s_o\right] \ln\left(1 - \frac{p}{P_{eq}}\right)
$$

(3)

where p is the concentration of product at time t, p_{eq} is the equilibrium concentration of product, and s_o is the initial concentration of substrate. It is assumed in deriving Equation (3) that the initial concentration of product is zero. If some product is present initially, the complete expression is given by:

$$
\left[\frac{V_s}{K_s} + \frac{V_p}{K_p}\right] t = \left[\frac{1}{K_s} - \frac{1}{K_p}\right] (p - p_o) - \left[1 + \frac{m_o}{K_p} + \frac{\left(\frac{1}{K_s} - \frac{1}{K_p}\right)}{\left(\frac{V_s}{K_s} + \frac{V_p}{K_p}\right)}\left(\frac{V_p}{K_p}\right) m_o\right] \ln \frac{(P_{eq} - p)}{(P_{eq} - P_o)}
$$

(4)

where p_o is the initial concentration of product, and m_o is the sum of the initial concentrations of substrate and product.

Equation (3), which is an integral rate equation, can be transformed into three linear forms corresponding to the Lineweaver-Burk, Eadie-Hofstee, and Woolf linear forms for the initial rate equation (Alberty, 1959).

If enzyme is not stable and inactivation follows a first-order rate law (Reiner, 1964; Wang and Humphrey, 1969; Ho and Humphrey, 1970) of the form:

$$
\frac{de_T}{dt} = -ke_T
$$

(5)

where k is a first-order rate constant and e_T is the total concentration of active enzyme in its uncomplexed and complexed forms, then Equation (1) can be combined with Equation (5) and integrated to obtain an expression for the time course of a batch isomerization employing enzyme subject to first-order inactivation. This expression is identical to equation (4), except t is replaced by:

$$
\frac{1 - \exp(-kt)}{k}
$$

In general, initial rate expressions like Equation (1) have usually been applied for studying reactions occurring in dilute solutions, although there are some cases where similar expressions have been applied to reactions occurring in concentrated solutions (e.g., hydrolysis of invertase - see Bowski, et al., 1971). In the case of industrial isomerization of glucose to fructose, it is essential to run reactions in concentrated sugar solutions. For example, in an early paper Takasaki (1966) presented integral rate data for batch isomerizations of sugar solutions ranging in concentration from 18 to 70 wt. %. In such concentrated solutions, simple rate laws like Equation (1) may no longer hold (Taraszka and Alberty, 1964). For example, in some cases it may be necessary to replace solute concentrations with thermodynamic activities if rate constants are to be independent of substrate and product concentrations. Such manipulations, however, are undoubtedly too lavish in the industrial context of glucose isomerization. Here the sound rule cited by James Wei (1966) comes into play. Known as the "Principle of Optimum Sloppiness," this simple rule states that "...one should use all the precision in the world where it is needed, but be sloppy where great effort is not required and would be better spent elsewhere." In the case of glucose isomerization, that effort probably should be applied to such problems as enzyme production, enzyme stability, and composition of the glucose/fructose product.

In fact, Havewala and Pitcher (1974) found that a radically simplified form of Equation (4) is adequate for fitting data they obtained with batch reactors employing isomerase immobilized on coated glass beads, and also with continuous column reactors again containing isomerase on coated glass beads over which 50% dextrose solutions were run. Specifically, these authors found that simple plots of their kinetic data were linear, as shown in Figure 10. (In Figure 10, X is the fraction of sugar [glucose and fructose] present as fructose at time t, X_i is the fraction present as fructose initially, and X_e is the fraction present as fructose at equilibrium.) Such linearity is predicted by Equation (4) if the first term on the right-hand-side of that expression can be neglected, e.g., if $K_s \approx K_p$. If $K_s = K_p \equiv K_m$, then Equation (1) simplifies to:

$$v = k_1 s - k_{-1} p \qquad (6)$$

where

$$k_1 \equiv \frac{V_s}{K_m + m_o}$$

$$k_{-1} \equiv \frac{V_p}{K_m + m_o}$$

Table X

Immobilized Isomerase and Isomerase Reactors

Carrier	Method of Immobilization	Activity Density *	Reactor Type	Reference
I. Cell-fixed Isomerase				
A. Cells free in suspension or packed in columns				
(1) S. albus	heat-fixing	-------------	continuous packed bed; batch	Takasaki, et al. (1969b, 1970)
(2) S. species (ATCC 21175)	heat-fixing	-------------	batch	Cotter, et al. (1971)
(3) S. species (ATCC 21175)	heat-fixing	110 units/gm dried cell-fixed isomerase (60°C)	continuous packed bed	Lloyd, et al. (1972)
(4) S. species (ATCC 21175)	heat-fixing and treatment with Variquot 415	200 units/gm dried cell-fixed isomerase	-------------	Dworschack, et al. (1973a)
B. Immobilized cells				
(1) collagen	entrapment	12.1 units/gm complex (70°C)	continuous packed bed (spiral wound membrane)	Vieth, et al. (1973)
(2) cellulose acetates	entrapment	-------------	continuous packed bed	Kolarik, et al. (1974)
II. Immobilized Cell-free Isomerase				
(1) DEAE-Sephadex	adsorption	-------------	continuous packed bed	Tsumura and Ishikawa (1967)
(2) polyacrylamide	entrapment	-------------	batch	Strandberg and Smiley (1971)
(3) glass beads	diazotization of amino-arysilane derivative of porous glass	1.0-3.0 units/gm glass (60°C)	batch	Strandberg and Smiley (1972)
(4) DEAE-cellulose (5) ECTEOLA-cellulose	adsorption	-------------	batch	Sipos (1973)
(6) collagen	electrocodeposition	13 units/cc complex	continuous packed bed (spiral wound membrane)	Weng and Vieth (1973)
(7) cellulose triacetate	entrapment	370-550 units/gm polymer (70°C)	batch	Giovenco, et al. (1973); Dinelli, et al. (1973).
(8) zirconia coated porous glass	glutaraldehyde coupling to silanized coated porous glass	700-900 units/gm glass (60°C)	continuous packed bed; batch	Havewala and Pitcher (1974).
(9) phenol-formaldehyde resins	adsorption	-------------	-------------	Olson and Stanley (1974).

*enzyme unit = 1.0 μmole fructose/min at conditions specified in reference (when available, temperature is indicated in parentheses).

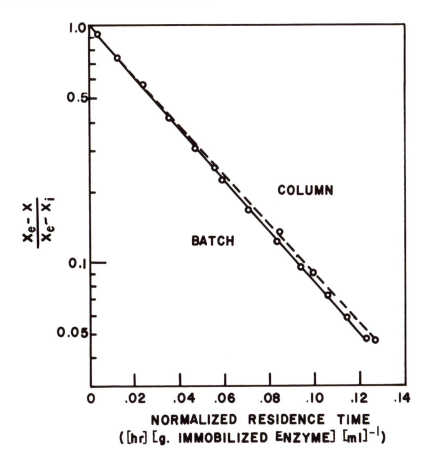

Figure 10: Kinetic data of Havewala and Pitcher (1974) plotted on coordinates to test for compatibility with quasi-first-order reversible kinetics. Data for both batch and continuous column reactors employing isomerase immobilized on coated glass are shown.

Equation (6) almost has the form of a simple reversible first-order kinetic law, except that the first-order rate constants, k_1 and k_{-1}, depend on initial sugar concentration, m_0. Therefore, Equation (6) is a quasi-first-order kinetic law.

   The variation of reaction rate with temperature, as reported by Havewala and Pitcher (1974), is shown in Figure 11 for soluble isomerase and isomerase immobilized on coated glass beads. The activation energies for the two cases are indistinguishable, about 15 kcal/mole.

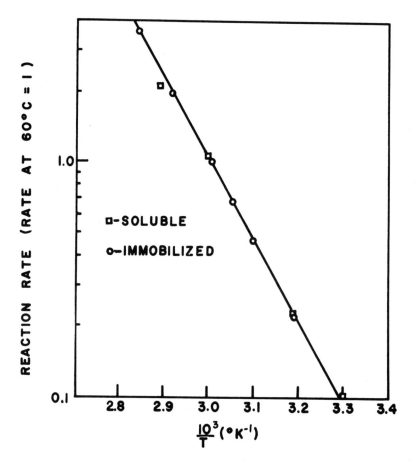

Figure 11: Arrhenius plot for soluble and immobilized glucose isomerase (Havewala and Pitcher, 1974).

The variations of reaction rate with pH and inorganic ion concentrations have been discussed by several workers. As already mentioned, arsenate is not required by glucose/xylose isomerase, but it is required for glucose-isomerizing activity of glucose-6-phosphate isomerase. Takasaki, et al. (1969a and 1969b) found that cobalt and magnesium ions are activators for (xylose) isomerase recovered from Streptomyces albus, and that the pH optimum is 8-8.5. Yamanaka (1968) found that manganese ion is an activator for isomerase recovered from Lactobacillus brevis, and that the pH optimum is 6-7 (Yamanaka, 1963b). Danno (1970b) found that cobalt, magnesium, and manganese ions are activators for isomerase recovered from Bacillus coagulans, and that the pH optimum is 7-7.5.

Immobilization. Table X summarizes methods by which glucose
isomerase has been fixed to cells or immobilized in the cell-free
form. The objectives of this work include: (1) stabilization of
isomerase activity for as long as possible; (2) minimization of
enzyme activity loss resulting from immobilization procedures;
(3) packaging of isomerase in a versatile module so that it can
be employed in a practical reactor configuration for efficient
reuse; and (4) attainment of high enzyme loadings (units/gm support),
but not beyond the point where diffusional restriction of catalytic
effectiveness becomes significant.

As discussed above, glucose isomerase fixed to cells immobilized in collagen and used in a packed column (Vieth, et al.,
1973) had the best apparent stability of all reported preparations.
As also already mentioned, however, it was not made clear to what
extent isomerase was really stabilized because of the possibility
of diffusional and microenvironmental influences "disguising"
actual instability. Additionally, sugar concentrations employed
for stability tests were relatively low (1.0M). On the other hand,
the system described by Havewala and Pitcher (1974), employing
isomerase immobilized on coated porous glass beads and packed
in a column, had an apparent stability of the same order as the
system of Vieth et al.,(1973), and evidence was presented to show
that no diffusional influences to disguise isomerase instability
were present. In addition, relatively high sugar concentrations
(50 wt.%) were employed. Unfortunately, information for cost
comparisons are not available for most immobilized enzyme
preparations.

Relatively low losses of isomerase activity as a result of
the immobilization procedure were encountered by Havewala and
Pitcher (1974). About 40% of the total enzyme activity employed
during immobilization remained present in the final preparation.
Strandberg and Smiley (1972) found that at most, only about 30%
of the activity they introduced for immobilization on uncoated
porous glass was measured in the recovered preparation. Only
about 15% was measured after entrapment in polyacrylamide
(Strandberg and Smiley, 1971), but much of the apparent activity
loss in this case might have been due to diffusional influences.
As already pointed out, Giovenco, et al., (1973) claimed that no
isomerase activity was lost upon immobilization in spun cellulose
triacetate fibers, but only 60% of the activity was measured,
probably because of diffusional limitations. As also already
mentioned, Takasaki, et al. (1969b) claimed that 80-90% of the
total glucose isomerase present in a cell culture could be cell-
fixed. Similarly, Lloyd, et al. (1972) claimed that a loss of
total cellular enzymatic activity as small as 15% following a
20 hour extraction of heat-fixed cells could be attained.

Table X also contains a tabulation of the reported densities
of immobilized isomerase activity. The highest values are for
covalent immobilization of partially purified enzyme on coated
porous glass (Havewala and Pitcher, 1974). This is perhaps
attributable to the high specific surface area of this support
and the uniform accessibility of its interior structure.

Immobilized Enzyme Kinetics. The observed kinetics of immobilized enzymes may behave differently from those of enzymes
free in solution because of diffusional and electrical effects
(see Hamilton, et al., 1973 and 1974b, and references cited therein),
as well as changes in enzyme conformation. Since glucose is an
uncharged substrate, it should be neither electrically attracted
nor repelled from a charged carrier. However, electrical effects
with a charged support might cause the pH at the site of action of
immobilized isomerase to be different from bulk pH, and so the pH
activity profile might be different from that for isomerase free
in solution. Diffusional effects might cause immobilized isomerase
to act at less than maximum catalytic efficiency, especially if
isomerase is embedded within porous supports. However, Havewala
and Pitcher (1974) found no evidence of diffusional limitations
with isomerase immobilized on coated porous glass beads. On the
other hand, Kolarik, et al. (1974) did find evidence of diffusional
limitations in studies they have begun on isomerase fixed in cells
which are in turn entrapped in cellulose acetates.

Reactor design and enzyme reuse. Table X lists reactor
configurations used by various workers for enzymatic isomerization
of glucose to fructose. Basically, the configurations employed
included batch and continuous tank reactors and continuous packed
bed reactors. In cases where isomerase was not used continuously
in a packed bed, it was immobilized in some form so that recovery
by filtration or centrifugation was possible. We have found no
reports of ultrafiltration reactors of the type described in
general by Butterworth, et al. (1970), although membrane filtration
has been used to recover glucose isomerase from reaction mixtures
(Barton and Denault, 1971).

Three types of reactor processing problems have been reported
on the industrial scale: (1) color development during isomerization,
(2) pH control during isomerization, and (3) pressure drop across
beds packed with cell-fixed isomerase. Cotter, et al. (1971)
found that incorporation of a water soluble salt of sulfurous acid
(e.g., a sulfite or bisulfite) into the isomerization liquor
reduces color formation (and also increases enzyme stability) as
does short residence times when highly active isomerase preparations are employed (Havewala and Pitcher, 1974). Observed pH drops
during isomerization are due to limited oxidation of sugar to

organic acids and can be controlled by addition of a pH regulator, such as calcium carbonate, magnesium carbonate, or an ion exchange resin (Takasaki and Kamibayashi, 1973). Pressure drop across beds packed with cell-fixed isomerase can be a problem because cells form compressible beds. Lloyd, et al. (1972) reported dealing with this problem by using shallow beds of cells contained within a pressure leaf filter apparatus. If isomerase immobilized on porous glass beads is used in a packed bed, pressure drop is very low (Havewala and Pitcher, 1974). Heat of reaction is endothermic and small, and apparently temperature control is not a problem (Havewala and Pitcher, 1974).

Alternative isomerase reactor configurations have been discussed by Havewala and Pitcher (1974). As they point out, a continuous stirred tank reactor is less efficient than a plug flow reactor for the pseudo-first-order reversible kinetics these authors found adequately fitted their rate data, and labor costs are higher for a batch reactor. These authors performed some design calculations based on a system using multiple continuous columns containing isomerase on coated porous glass. Isomerase inactivation was taken into account. They summarize:

"To process an average of 10 million pounds of 50% glucose solution per year at 60°C (45% conversion to fructose) eleven columns 6 inches in diameter and 3 feet high containing immobilized enzyme with 800 IGIU/g activity [IGIU = International Glucose Isomerase Units 800 is a figure typically obtained experimentally by these authors] are necessary. In contrast the comparable batch process at 50°C as calculated from data of Cotter et al. (1971) would require processing 2500 cu. ft. of glucose solution per batch, over 300 times the total immobilized enzyme bed volume. The amount of enzyme used in this batch process is approximately 10 times that required by the immobilized enzyme column approach assuming 2 day enzyme half-life and 3 half-life utilization at 60°C. This calculation takes into account the fact that only 50 to 60% of the enzyme bound is active, possibly due to binding to active sites. The estimated pressure drops of less than 10 psi for column operation should provide no problem for immobilized enzyme usage. The longer residence time (3 to 4 days) for the batch process results in product discoloration not observed with the shorter residence times (less than one hour) of the immobilized enzyme system. Also enzyme must be removed from the product of the batch process."

Product Purification and Recovery

After enzymatic isomerization, high-fructose corn syrups are refined by operations traditional to the sugar industry: filtration, carbon decolorization, ion exchange deionization, and evaporation to concentrate to desired solids content (Kooi and Smith, 1972; Newton and Wardrip, 1973).

Production of pure crystalline fructose presents more of a problem. First, a means for separation of fructose from glucose is required. Three different procedures effecting such a separation have been described in the literature:

1. Oxidation of glucose using glucose oxidase, followed by precipitation
2. Lime precipitation
3. Column chromatography

Oxidation of glucose using glucose oxidase, followed by precipitation of gluconate with fructose remaining in solution, is described by Holstein and Holsing (1962). This method is practiced by Dawe's Laboratories, Inc., and has been the basis for crystalline fructose production in the United States for the past several years (Ward, 1967). An obvious inconvenience as a means for high-volume, low-cost production of fructose is the associated production of large quantities of gluconate by-product.

Fructose may also be separated from glucose using the lime process developed by the National Bureau of Standards (Jackson, et al., 1925; Haack, et al., 1964), in which calcium-fructosate is precipitated with glucose remaining in solution. According to Ward (1967), this method was tedious and uneconomical when practiced on an industrial scale several years ago and has been abandoned. More recently, a modified calcium-fructosate precipitation process has been patented by Hara and Kazuo (1972).

A third method for separation of fructose from glucose involves column chromatographic procedures using a packing with which fructose preferentially interacts, e.g., a sulfonated polystyrene cationic resin partially in a salt (calcium, barium, strontium, silver) form (Serbia and Aguirre, 1962; Lefevere, 1962a, 1962b; Boehringer & Sons, 1967; Lauer, et al., 1969), or an anionic resin in the bisulfite form (Lindberg and Slessor, 1967; Takasaki, 1972). The design of large-scale chromatography columns for low cost processing has been described (Timmins, et al., 1969), and there have been several reports (Takasaki, 1972; Goldsmith, 1973) of current industrial application of this technique for separating fructose from invert sugar or glucose-isomerized syrup.

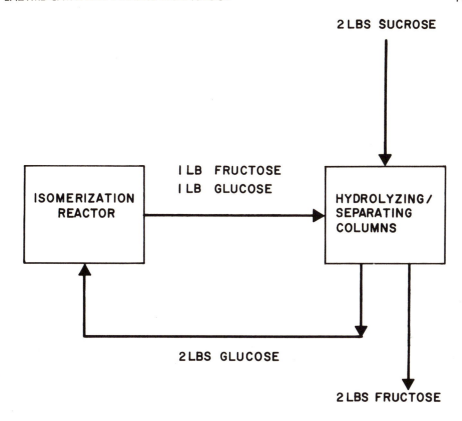

Figure 12: Coupling of glucose isomerase process and the Boehringer process to produce two pounds of fructose for every two pounds of sucrose fed.

Once fructose has been separated from glucose, it can be recovered in solid form by adjusting to the proper concentration and pH and then crystallizing from aqueous solution (Jackson et al., 1925; Young and Jones, 1952; Boehringer & Sons, 1968), or else by adding methanol (Holstein and Holsing, 1962) or ethanol (Barry and Honeyman, 1952; Verstraeten, 1967).

Possible Future Processing Objectives

The search for genetically improved microbial producers of glucose isomerase will undoubtedly continue, as will attempts to manipulate environmental factors to raise in vivo isomerase levels. Immobilization efforts underway will probably continue in the hope

of finding a low-cost, mechanically satisfactory carrier which can
be used to improve isomerase stability under reaction conditions.
Additionally, use of isomerase immobilized on non-cellular
carriers eliminates much extraneous cellular material, which in
turn decreases color development problems during isomerization and
permits less extensive refining operations.

Efforts to increase conversion of glucose to fructose, perhaps
by using complexing agents, will also probably continue. If
enough market incentive develops, separation of unconverted
glucose from fructose followed by fructose crystallization may
become attractive. Moreover, it might prove useful to incorporate
isomerase into the Boehringer process to enable complete conversion of sucrose to fructose (see Figure 12).

Acknowledgement

We gratefully acknowledge support in part by NSF Grant
GI-34284. Bruce Hamilton is a recipient of a Pillsbury Fellowship
(administered by the Institute of Food Technologists) for which
he is grateful.

REFERENCES

Alberty, R.A., "The Rate Equation for an Enzymic Reaction," in
The Enzymes (edited by P.D. Boyer, H. Lardy, and K. Myrbäck),
2nd Edition, page 143 (1959)

Archer, M.C., Ragnarsson, J.O., Tannenbaum, S.R., and Wang, D.I.C.,
"Enzymatic Solubilization of an Insoluble Substrate, Fish Protein
Concentrate: Process and Kinetic Considerations," Biotech.
Bioeng., 15, 181 (1973)

Barfoed, H., "Enzymatic De-sizing of Textiles," Process Biochemistry,
17 (August 1970)

Barker, S.A., Hatt, B.W., and Somers, P.J., "The Effect of
Areneboronic Acids on the Alkaline Conversion of D-Glucose into
D-Fructose," Carbohydrate Research, 26, 41 (1973a)

Barker, S.A., Hatt, B.W., Somers, P.J., and Woodbury, R.R., "The
Use of Poly(4-vinylbenzeneboronic acid) Resins in the Fractionation
and Interconversion of Carbohydrates," Carbohydrate Research, 26,
55 (1973b)

Barry, C.P., and Honeyman, J., "Fructose and its Derivatives," in Adv. Carbohydrate Chem. (edited by C.S. Hudson, M.L. Wolfrom, and S.M. Cantor), Vol. 7, Academic Press, page 53 (1952)

Barton, R.R., and Denault, L.J. (to Miles Laboratories Inc.), "Recovery of Enzymes," German Patent 2,031,258 (Jan. 7, 1971)

deBecze, G.I., "Enzymes, Industrial Applications," in Encyclopedia of Polymer Science and Technology, Vol. 6, Interscience (1967)

Beckhorn, E.J., "Enzyme, Industrial Production of," in Encyclopedia of Science & Technology, 3rd Ed., Vol. 5, McGraw-Hill (1971)

Bengtson, B.L., and Lamm, W.R. (to Standard Brands Inc.),"Process for Isomerizing Glucose to Fructose," U.S. Patent 3,654,080 (April 4, 1972)

Berman, G.A., and Murashige, K.H., "Synthetic Carbohydrate," The Stanford-Ames NASA/ASEE Summer Faculty Systems Design Workshop, Final Report (1973)

Bernath, F.R., and Vieth, W.R., "Collagen as a Carrier for Enzymes: Materials Science and Process Engineering Aspects of Enzyme Engineering," this volume (1974)

Birch, G.G., Green, L.F., and Coulson, C.B. (editors), Glucose Syrups, Elsevier (1970)

Boehringer, C.F., and Sons, "A Process for Obtaining Pure Glucose and Fructose," British Patent 1,085,696 (Oct. 4, 1967)

Boehringer, C.F., and Sons, "Process for the Production of Anhydrous Crystalline Fructose," British Patent 1,117,903 (June 26, 1968)

Bowski, L., Saini, R., Ryu, D.Y., and Vieth, W.R., "Kinetic Modeling of the Hydrolysis of Sucrose by Invertase," Biotech. Bioeng., 13, 641 (1971)

Brownewell, C.E., and Streets, B.W.(to Miles Laboratories Inc.), "Glucose Isomerase," German Patent 2,219,713 (Oct. 26, 1972)

Butterworth, T.A., Wang, D.I.C., and Sinskey,A.J., "Application of Ultrafiltration for Enzyme Retention During Continuous Enzymatic Reaction," Biotech. Bioeng., 12, 615 (1970)

Carbonell, R.G., and Kostin, M.D., "Enzyme Kinetics and Engineering," AIChE J., 18, 1 (1972)

Chang, T.M.S., Artificial Cells, Charles C. Thomas Publisher (1972)

Chem. Eng. News, "Patent issues on cure for overdose of alcohol," 56 (July 5, 1971a)

Chem. Eng. News, "Fructose patent queried," 68 (Oct. 18, 1971b)

Chem. Eng. News, "Corn syrup from immobilized enzymes," 36 (April 23, 1973)

Clement, M.T., and Gibbons, W.E., "Aeromonas hydrophilia (Pseudomonas hydrophila) NRC 491 and 492 Established as Aerobacter cloacae," Can. J. Microbiol., 6, 591 (1960)

Conover, L.H., "Discovery of Drugs from Microbiological Sources," in Drug Discovery (edited by B. Bloom and G.E. Ollyot), ACS (1971)

Cooney, C.L., Weaver, J.C., Tannenbaum, S.R., Faller, D.V., Shields, A., and Jahnke, M., "The Thermal Enzyme Probe -- A Novel Approach to Chemical Analysis," in Enzyme Engineering II (edited by E.K. Pye and L.B. Wingard), Plenum Press (1974)

Cotter, W.P., Lloyd, N.E., and Hinman, C.W. (to Standard Brands, Inc.), "Method for Isomerizing Glucose Syrups," U.S. Patent 3,623,953 (Nov. 30, 1971)

Dahlqvist, A., Mattiasson, B., and Mosbach, K., "Hydrolysis of B-Galactosides Using Polymer-Entrapped Lactose. A Study Towards Producing Lactose-Free Milk," Biotech. Bioeng., 15, 395 (1973)

Danno, G., "Studies on D-Glucose-Isomerizing Enzyme from Bacillus coagulans, Strain HN-68: Part IV. Purification, Crystallization and Some Physicochemical Properties," Agr. Biol. Chem., 34, 1795 (1970a)

Danno, G., "Studies on D-Glucose-Isomerizing Enzyme from Bacillus coagulans, Strain HN-68: Part V. Comparative Study of the Three Activities of D-Glucose, D-Xylose and D-Ribose Isomerization of the Crystalline Enzyme," Agr. Biol. Chem., 34, 1805 (1970b)

Dean, A.C.R., "Influence of Environment on the Control of Enzyme Synthesis," J. Appl. Chem. Biotechnol., 22, 245 (1972)

Demain, A.L., "Oversynthesis of Microbial Enzymes," in Developments in Industrial Microbiology, 12, 56 (1971a)

Demain, A.L., "Overproduction of Microbial Metabolites and Enzymes Due to Alteration of Regulation," in Adv. Biochem. Eng., 1, 113 (1971b)

Demain, A.L., "Theoretical and Applied Aspects of Enzyme Regulation and Biosynthesis in Microbial Cells," in Enzyme Engineering (edited by L.B. Wingard), Interscience, page 21 (1972)

Dinelli, D., Morisi, F., Giovenco, S., and Pansolli, P. (to Snam Progetti, Rome), "Enzymic Manufacture of Fructose and Sirups Containing Glucose and Fructose," German Patent 2,303,872 (Aug. 16, 1973)

Doig, A.R., "Stability of Enzymes from Thermophillic Microorganisms," in Enzyme Engineering II (edited by E.K. Pye and L.B. Wingard), Plenum Press (1974)

Dunnill, P., in Fermentation and Enzyme Technology, Wiley (1974)

Dworschack, R.G., and Lamm, W.R. (to Standard Brands Inc.), "Process for Growing Microorganisms," U.S. Patent 3,666,628 (May 30, 1972)

Dworschack, R.G., Chen, J.C., Khwaja, A., and White, W.H. (to Standard Brands Inc.), "Stabilization of Enzymes," German Patent 2,251,855 (April 26, 1973a)

Dworschack, R.G., Chen, J.C., Lamm, W.R., and Davis, L.G. (to Standard Brands Inc.), "Growing Streptomyces," U.S. Patent 3,736,232 (May 29, 1973b)

Faith, W.T., Neubeck, C.E., and Reese, E.T., "Production and Application of Enzymes," in Adv. Biochem. Eng., 1, 77 (1971)

Fetzer, W.R., and Evans, J.W. (to Union Starch & Refining Company), "Extra Sweet Corn Syrup," U.S. Patent 2,487,121 (Nov. 8, 1949)

Gardner, C.R., Colton, C.K., Langer, R.S., Hamilton, B.K., Archer, M.C., and Whitesides, G.M., "Enzymatic Regeneration of ATP from AMP and ADP. Part I. Thermodynamics, Kinetics, and Process Development," Enzyme Engineering II (edited by E.K. Pye and L.B. Wingard), Plenum Press (1974)

Giovenco, S., Morisi, F., and Pansolli, P., "Properties of Free and Immobilized Glucose Isomerase," FEBS Letters, 36, 57 (1973)

Goldsmith, R.L., Abcor Inc., personal communication (1973)

Gordon, H.W. (to Julius Schmid, Inc.), "Oral Administration of Fructose for Alcohol Intoxication," U.S. Patent 3,584,122 (June 8, 1971)

Guilbault, G.G., _Enzymatic Methods of Analysis_, Pergamon Press (1970)

Haack, E., Braun, F., and Kohler, K. (to C.F. Boehringer & Sons), "D-Fructose," German Patent 1,163,307 (Feb. 20, 1964)

Haldane, J.B.S., _Enzymes_, reprinted by M.I.T. Press, 1965 (1930)

Hamilton, B.K., Stockmeyer, L.J., and Colton, C.K., "Comments on Diffusive and Electrostatic Effects with Immobilized Enzymes," J. Theor. Biol., 41, 547 (1973)

Hamilton, B.K., Montgomery, J. P., and Wang, D.I.C., "Application of Enzyme Reactions for Preparative-Scale Synthesis," in _Enzyme Engineering II_ (edited by E.K. Pye and L.B. Wingard), Plenum Press (1974a)

Hamilton, B.K., Gardner, C.R., and Colton, C.K., "Basic Concepts in the Effects of Mass Transfer on Immobilized Enzyme Kinetics," this volume (1974b)

Hara, K., and Kazuo, S., (to Nippon Yanoh Inc.), "Fructose Separation from Invert Sugar," Japanese Patent 27,942(72) (July 25, 1972)

Harden, J.D., "On-Line Control Optimizes Processing: Part I," Food Engineering, 59 (Dec. 1972)

Harden, J.D., "On-Line Control Optimizes Processing: Part II," Food Engineering, 65 (Jan. 1973)

Havewala, N.B., and Pitcher, W.H., "Immobilized Glucose Isomerase for the Production of High Fructose Syrups," in _Enzyme Engineering II_ (edited by E.K. Pye and L.B. Wingard), Plenum Press (1974)

Heady, R. E., and Jacaway, W.A. (to CPC International Inc.), "Glucose Isomerase," German Patent 2,223,864 (Nov. 30, 1972)

Ho, L.Y., and Humphrey, A.E., "Optimal Control of an Enzyme Reaction Subject to Enzyme Deactivation. I. Batch Process," Biotech. Bioeng., 12, 291 (1970)

Holstein, A.G., and Holsing, G.C. (to Dawes Laboratories, Inc.), "Method for the Production of Levulose," U.S. Patent 3,050,444 (Aug. 21, 1962)

Horwath, R.O., and Cole, G.W. (to Standard Brands Inc.), "Glucose Isomerase," German Patent 2,247,922 (April 19, 1973)

Inman, D.J., and Hornby, W.E., "The Immobilization of Enzymes on Nylon Structures and Their Use in Automated Analysis," Biochem. J., 129, 255 (1972)

Jackson, R.F., Silsbee, C.G., and Proffitt, M.J., "The Preparation of Levulose," Scientific Papers of the Bureau of Standards, 20, 587 (1924-1926)

Jensen, G., "Bacillus Derived Detergent Enzymes," Process Biochemistry, 23 (August 1972)

Kainuma, K., and Suzuki, S., "Isomerization of Dextrose into Fructose. I. Isomerization with Sodium Hydroxide," Nippon Nogei Kagaku Kaishi, 38 (12), 556 (1964)

Kainuma, K., Tadokoro, K., and Suzuki, S., "Isomerization of Dextrose into Fructose. II. Isomerization with Various Types of Alkaline Reagents," Nippon Nogei Kagaku Kaishi, 40 (1), 35 (1966a)

Kainuma, K., Tadokoro, K., and Suzuki, S., "Isomerization of Dextrose into Fructose. III. Effects of Cations on the Isomerization of Dextrose," Nippon Nogei Kagaku Kaishi, 42 (4), 173 (1968a)

Kainuma, K., Yamamoto, K., and Suzuki, S., "Isomerization of Dextrose into Fructose. IV. Design, Construction, and Operating Conditions of a Pilot Plant for Continuous Isomerization," Nippon Nogei Kagaku Kaishi, 42 (5), 243 (1968b)

Kainuma, K., Yamamoto, K., Tadokoro, K., and Suzuki, S., "Isomerization of Dextrose into Fructose. V. Isomerization by Continuous Flow System Under Pilot Plant Scale," Nippon Nogei Kagaku Kaishi, 42 (5), 249 (1968c)

Kainuma, K., and Suzuki, S., "Isomerization of Dextrose into Fructose by the Alkali Method," Staerke, 18 (5), 135 (1966b)

Kainuma, K., and Suzuki, S., "Isomerization of Dextrose into Fructose by the Alkali Method. II. Design and Installation of the Pilot Plant for the Continuous Isomerization and Determination of its Operating Conditions," Staerke, 19 (3), 60 (1967a)

Kainuma, K., and Suzuki, S., "Isomerization of Dextrose into Fructose by the Alkali Method. III. Isomerization with the Pilot Plant of Continuous Flow System," Staerke, 19 (3), 66 (1967b)

Katz, E., Ehrenthal, I., and Scallet, B.L. (to Anheuser-Busch Inc.), "Process of Making High D.E. Fructose Containing Syrups," U.S. Patent 3,690,948 (Sept. 12, 1972)

Kolarik, M.J., Chen, B.J., Emery, A.H., and Lim, H.C., "Glucose Isomerase Cells Entrapped in Cellulose Acetates," this volume (1974)

Kooi, E.R., and Smith, R.J., "Dextrose-Levulose Syrup from Dextrose," Food Technology, 57 (Sept. 1972)

Langlois, D.P., and Larson, R.F. (to A.E. Staley Manufacturing Company), "Interconversion of Sugars Using Anion Exchange Resins," U.S. Patent 2,746,889 (May 22, 1956)

Lantos, P.R., "What R&D Can Expect from Marketing Research," Chemtech, 588 (Oct. 1973)

Lauer, K., Weber, M., and Stoeck, G. (to C.F. Boehringer & Sons), "Method of Recovering Pure Glucose and Fructose from Sucrose or from Sucrose-containing Invert Syrups," U.S. Patent 3,483,031 (Dec. 9, 1969)

Lee, C.K., Hayes, L.E., and Long, M.E. (to R.J. Reynolds Tobacco Company), "Process of Preparing Glucose Isomerase," U.S. Patent 3,645,848 (Feb. 29, 1972)

Lefevre, L.J. (to Dow Chemical Company), "Separation of Fructose from Glucose Using Cation Exchange Resin Salts," U.S. Patent 3,044,905 (July 17, 1962a)

Lefevre, L.J. (to Dow Chemical Company), "Separation of Fructose from Glucose Using a Cation Exchange Resin Salt," U.S. Patent 3,044,906 (July 17, 1962b)

Lindberg, B., and Slessor, K.N., "Preparative Separations of Sugars on Bisulphite Resins," Carbohydrate Research, 5, 286 (1967)

Lloyd, N.E., Lewis, L.T., Logan, R.M., and Patel, D.N. (to Standard Brands, Inc.), "Process for Isomerizing Glucose to Fructose," U.S. Patent 3,694,314 (Sept. 26, 1972)

Marshall, D.L., Walter, J.L., and Falb, R.D., "Development of Techniques for the Insolubilization of Enzymes to Convert Glucose to Starch," Interim Report from Batelle Columbus Laboratories to NASA Ames Research Center (June 29, 1972)

Marshall, R.O. (to Corn Products Company), "Enzymatic Process," U.S. Patent 2,950,228 (Aug. 23, 1960)

Marshall, R.O., and Kooi, E.R., "Enzymatic Conversion of D-Glucose to D-Fructose," Science, 125, 648 (1957)

McLaren, A.D., and Packer, L., "Some Aspects of Enzyme Reactions in Heterogeneous Systems," in Adv. Enzymol. (edited by F.F. Nord), Vol. 33, page 245 (1970)

Mendicino, J.H., "Effect of Borate on the Alkali-catalyzed Isomerization of Sugars," J. Amer. Chem. Soc., 82, 4975 (1960)

Mosbach, K., and Larsson, P., "Preparation and Application of Polymer-Entrapped Enzymes and Microorganisms in Microbial Transformation Processes with Special Reference to Steroid 11-β-Hydroxylation and Δ^1-Dehydrogenation," Biotech. Bioeng., 12, 19 (1970)

Newton, J.M., and Wardrip, E.K., "High Fructose Corn Syrup," paper presented at ACS Spring Meeting, Dallas (1974)

Nicol, W.M., "Sweeteners in Foods," Process Biochem., 17 (Dec. 1971)

Noltmann, E.A., "Aldose-Ketose Isomerases," in The Enzymes (edited by P.D. Boyer), 3rd edition, Vol. 6, page 271 (1972)

Ollis, D.F., "Diffusion Influences in Denaturable Insolubilized Enzyme Catalysis," Biotech. Bioeng., 14, 871 (1972)

Ollis, D.F., Datta, R., Lieberman, R., and Saville, D.A., "Hydrolysis of Particulates by Immobilized Depolymerases," presented at 66th Annual A.I.Ch.E. Meeting, Philadelphia (1973)

Olson, A.C., and Stanley, W.L., "Use of Phenol Formaldehyde Resins and Glutaraldehyde to Immobilize Enzymes," this volume (1974)

Peller, L., and Alberty, R.A., "Multiple Intermediates in Steady State Enzyme Kinetics. I. The Mechanism Involving a Single Substrate and Product," J. Amer. Chem. Soc., 81, 5907 (1959)

Reed, G., Enzymes in Food Processing, Academic Press (1966)

Reiner, J.M., "Quantitative Aspects of Enzymes and Enzyme Systems," Chapter V in Comprehensive Biochemistry (edited by M. Florkin and E.H. Stolz), Vol. 12, Elsevier (1964)

Rubin, D.H., "A Technology Assessment Methodology: Enzymes, Industrial," The MITRE Corp. (June, 1971)

Sato, T., and Tsumura, S. (to National Institute of Food Sciences, Tokyo), "Manufacture of Fructose from Glucose by Use of Microorganisms," Japanese Patent 17,640 ('66) (Oct. 7, 1966)

Scallet, B.L., and Ehrenthal, I. (to Anheuser-Busch Inc.), "High D.E. Corn Type Starch Conversion Syrup and Methods of Making Same," U.S. Patent 3,305,395 (Feb. 21, 1967)

Scallet, B.L., and Ehrenthal, I. (to Anheuser-Busch Inc.), "Process of Purifying High D.E. - Very Sweet Syrups," U.S. Patent 3,383,245 (May 14, 1968)

Scallet, B.L., Katz, E., and Ehrenthal, I. (to Anheuser-Busch Inc.), "Process of Making High D.E. Fructose Containing Syrups," U.S. Patent 3,690,948 (Sept. 12, 1972)

Schnyder, B.J., "Letter to the Editor," Enzyme Technology Digest, 1, 165 (1973)

Serbia, G.R., and Aguirre, P.R. (to Central Aguirre Sugar Company), "Separation of Dextrose and Levulose," U.S. Patent 3,044,904 (July 17, 1962)

Sherwood, M.A., "Enzymes in Industry," Process Biochemistry, 279 (August 1966)

Sipos, T. (to Baxter Laboratories Inc.), "Syrup Conversion with Immobilized Glucose Isomerase," U.S. Patent 3,708,397 (Jan. 2, 1973)

Sizer, I.W., "Medical Applications of Microbial Enzymes," Adv. App. Micro, 15, 1 (1972)

Slote, L., "Development of Immobilized Enzyme Systems for Enhancement of Biological Waste Treatment Processes," report to Water Quality Office, EPA (July, 1970)

Smolarsky, M., Berger, A., Kurn, N., and Bosshard, H.R., "A New Method for the Synthesis of Optically Active α-Amino Acids and Their N^{α} Derivatives via Acylamino Malonates," J. Org. Chem., 38, 457 (1973)

Strandberg, G.W., and Smiley, K.L., "Free and Immobilized Glucose Isomerase from Streptomyces phaeochromogenes," App. Microbiol., 21, 588 (1971)

Strandberg, G.W., and Smiley, K.L., "Glucose Isomerase Covalently Bound to Porous Glass Beads," Biotech. and Bioeng., 14, 509 (1972)

Takasaki, Y., "Studies on Sugar-isomerizing Enzyme: Production and Utilization of Glucose Isomerase from Streptomyces Sp.," Agr. Biol. Chem., 30, 1247 (1966)

Takasaki, Y., "Kinetic and Equilibrium Studies on D-Glucose-D-Fructose Isomerization Catalyzed by Glucose Isomerase from Streptomyces Sp.," Agr. Biol. Chem., 31, 309 (1967)

Takasaki, Y., "Studies on Sugar-isomerizing Enzymes: Effect of Borate on Glucose-fructose Isomerization Catalyzed by Glucose Isomerase," Agr. Biol. Chem., 35, 1371 (1971)

Takasaki, Y., "On the Separation of Sugars," Agr. Biol. Chem., 36, 2575 (1972)

Takasaki, Y. (to Institute for Fermentation, Osaka), "Production of Glucose Isomerase," Japanese Patent 49,981 ('73) (July 14, 1973)

Takasaki, Y., and Kambayashi, A. (to Bureau of Industrial Technology, Japan), "Glucose Isomerase Production," Japanese Patent 19,086 ('72) (Sept. 19, 1972)

Takasaki, Y., and Kamibayashi, A. (to the Agency of Industrial Science and Technology Governmental, Tokyo), "Enzymatic Method for Manufacture of Fructose from Glucose," U.S. Patent 3,715,276 (Feb. 6, 1973)

Takasaki, Y., Kosugi, Y., and Kanbayashi, A., "Studies on Sugar-isomerizing Enzyme: Purification, Crystallization and Some Properties of Glucose Isomerase from Streptomyces sp.," Agr. Biol. Chem., 33, 1527 (1969a)

Takasaki, Y., Kosugi, Y., and Kanbayashi, A., "Streptomyces Glucose Isomerase," in Fermentation Advances (edited by D. Perlman), Academic Press, page 561 (1969b)

Takasaki, Y., and Tanabe, O. (to the Agency of Industrial Science and Technology, Tokyo), "Enzymatic Method for Converting Glucose in Glucose Syrups to Fructose," U.S. Patent 3,616,221 (Oct. 26, 1971)

Taraszka, M., and Alberty, R.A., "Extensions of the Steady-State Rate Law for the Fumarase Reaction," J. Phys. Chem., 68, 3368 (1964)

Tosa, T., Mori, T., Fuse, N., and Chibata, I., "Studies on Continuous Enzyme Reactions: Part V. Kinetics and Industrial Application of Aminoacylase Column for Continuous Resolution of Acyl-DL-Amino Acids," Agr. Biol. Chem., 33, 1047 (1969)

Tsumura, N., Isao, K., and Ishikawa, M., "Detection of Isomerases," Nippon Shokuhin Kogyo Gakkaishi, 14, 548 (1967)

Tsumura, N., and Ishikawa, M., "Continuous Isomerization of Glucose by a Column of Glucose Isomerase," Nippon Shokuhin Kogyo Gakkaishi, 14 (12), 539 (1967)

Tsumura, N., and Sato, T., "Conversion of D-Glucose to D-Fructose by a Strain of Soil Bacteria," Bull. Agric. Chem. Soc. Japan, 24, 326 (1960)

Tsumura, N., and Sato, T., "Enzymatic Conversion of D-Glucose to D-Fructose: Part I. Identification of Active Bacterial Strain and Confirmation of D-Fructose Formation," Agr. Biol. Chem., 25, 616 (1961)

Underkofler, L.A., "Enzymes," in Handbook of Food Additives (edited by T. E. Furia), CRC, page 51 (1968)

Van Dyke, B., Jr., Baddour, R.F., Bodman, S.W., and Colton, C.K., "Kinetics of Cellulose Enzymatic Hydrolysis," AIChE J. (in press)

Verstraeten, L.M.J., "D-Fructose and its Derivatives," in Adv. Carbohydrate Chem. (edited by M.L. Wolfrom and R.S. Tipson), Vol. 22, Academic Press, page 229 (1967)

Vieth, W.R., Wang, S.S., and Saini, R., "Immobilization of Whole Cells in a Membraneous Form," Biotech. Bioeng., 15, 565 (1973)

Vogels, R.J., "Enzymes in Washing Powders," Birmingham University Chemical Engineer, 58 (summer 1971)

Wang, D.I.C., and Humphrey, A.E., "Biochemical Engineering," Chem. Engng., 108 (Dec. 15, 1969)

Wang, S.S., and Vieth, W.R., "Collagen-Enzyme Complex Membranes and their Performance in Biocatalytic Modules," Biotech. Bioeng., 15, 93 (1973)

Warburton, D., Dunnill, P., Lilly, M.D., "Conversion of Benzyl-penicillin to 6-Aminopenicillanic Acid in a Batch Reactor and Continuous Feed Stirred Tank Reactor Using Immobilized Penicillin Amidase," Biotech. Bioeng., 15, 13 (1973)

Ward, G.E., "Production of Gluconic Acid, Glucose Oxidase, Fructose, and Sorbose," in Microbial Technology (edited by H. J. Peppler), Reinhold, page 200 (1967)

Weber, H.C., and Meissner, H.P., Thermodynamics for Chemical Engineers, 2nd edition, Wiley (1966)

Wei, J., "Disguished Kinetics," Ind. Eng. Chem., 58, 39 (1966)

Wieland, H., Enzymes in Food Processing and Products, Noyes Data Corporation, 1972

Wingard, L.B. (editor), Enzyme Engineering, Interscience (1972a)

Wiseman, A., "Industrial Enzyme Stabilisation," Process Biochem.,
14 (August 1973)

Wolnak, B., "Present and Future Technological and Commercial Status
of Enzymes," Study for the National Science Foundation, Washington,
D.C. (1972)

Yamanaka, K., "Sugar Isomerases: Part I. Production of D-Glucose
Isomerase from Heterolactic Acid Bacteria," Agr. Biol. Chem., 27,
265 (1963a)

Yamanaka, K., "Sugar Isomerases: Part II. Purification and
Properties of D-Glucose Isomerase from Lactobacillus brevis,"
Agr. Biol. Chem., 27, 271 (1963b)

Yamanaka, K., "Purification, Crystallization and Properties of the
D-Xylose Isomerase from Lactobacillus brevis," Biochem. Biophys.
Acta, 151, 670 (1968)

Yoshimura, S., Danno, G., and Natake, M., "Studies on D-Glucose
Isomerizing Activity of D-Xylose Grown cells from Bacillus coagulans,
Strain HN-68," Agr. Biol. Chem., 30, 1015 (1966)

Young, F.E., and Jones, F.T. (to U.S. Dept. of Agr.), "Levulose
Dihydrate," U.S. Patent 2,588,449 (March 11, 1952)

Zaborsky, O., Immobilized Enzymes, CRC Press (1973)

Zaborsky, O., "Enzyme Stabilization by Intramolecular Crosslinking,"
in Enzyme Engineering II (edited by E. K. Pye and L. B. Wingard),
Plenum Press (1974)

Zienty, M.F. (to Miles Laboratories Inc.), "Stabilization of Glucose
Isomerase in Streptomyces olivaceus Cells," German Patent 2,223,340
(November 23, 1972)

IMMOBILIZED α-AMYLASE FOR CLARIFICATION OF COLLOIDAL STARCH-CLAY SUSPENSIONS

K. L. Smiley, J. A. Boundy, B. T. Hofreiter, and
S. P. Rogovin
Northern Regional Research Laboratory, Agricultural
Research Service, U.S. Department of Agriculture,
Peoria, Illinois 61604

Lake and stream pollution remains a major national concern. New regulations, either now in effect or to be imposed in the next few years, will limit severely the amount of solids that industry and municipalities can discharge into natural waters. Among the industries affected are those either manufacturing starch or using it in processes where some starch may get into waste streams. Because dilute starch solutions suspend other solid materials, they constitute a problem in waste treatment.

The paper industry in particular has a serious problem with waste streams when some types of starch, such as unmodified, hydroxyethyl or oxidized, are present. In the manufacture of many specialty papers, starch or modified starch and often fillers, such as clay, are added to the pulp suspension to give the paper desired properties. A portion of this aqueous suspension, known as "white water," escapes from the process and after some clarification is discharged as waste. The solids in these effluent waters are difficult to remove even with the aid of flocculants. However, in amylase-treated white water they readily settle out with the aid of alum.

The use of soluble α-amylase (α-1,4-glucan 4-glucanohydrolase, EC 3.2.1.1) to degrade starch and its derivatives in paper mill effluent is costly because of the large volumes of solution required (Schwanke and Davis, 1973). It would be more practical to treat these starch solutions with an immobilized α-amylase. The enzyme would be used over long periods of time, significantly reducing its cost. This paper reports use of immobilized α-amylase on a laboratory scale to degrade starch in white water. Various methods of immobilization of α-amylase are evaluated.

MATERIALS AND METHODS

Source of Enzyme

Crystalline <u>Bacillus</u> <u>subtilis</u> α-amylase was purchased from Sigma Chemical Co., St. Louis, Mo. Crude bacterial α-amylase, designated HT-1000, was kindly supplied by Miles Chemical Co., Elkhart, Ind., and a similar product in liquid form, 86L, was donated by Rohm and Haas Co., Philadelphia, Pa.

Source of Starch

Penford gum, a dextrinized hydroxyethyl starch, is made by Penick & Ford, Ltd., Cedar Rapids, Iowa. Lintner starch was purchased from Pfanstiehl Chemical Corp., Kankakee, Ill. Mor-Rex, a corn dextrin, and pearl corn starch (Globe 3001) came from CPC International, Argo, Ill.

Source of Paper Mill Effluent

A base white water, pH 6.5, was collected from the tray of a pilot paper machine before starch addition. The machine furnish was a 50:50 blend of bleached hard- and softwood kraft pulp. Kaolin clay filler, rosin size, and alum were present in the furnish at levels of 20, 0.4, and 2.0% (based on dry pulp weight), respectively. The white water was passed over a 72 X 24 wire mesh screen to remove pulp and coarse solids. The water from a save-all served as a stock solution to prepare paper mill effluent containing cooked starches. Cooked starch was prepared in the laboratory by heating a 3% starch slurry at 85° C for 10 min. Stock white water was brought to the desired starch concentration by adding suitable amounts of cooked starch and stirring 15 min. Generally the amount of starch added was 0.01 or 0.02% based on solution weight. For some experiments, white water containing starch was collected directly from various locations at the pilot paper machine.

Attachment of α-Amylase to a Resinous Adsorbant

α-Amylase was attached to Duolite S-30 by the procedure of Olson and Stanley (1973). Diamond Shamrock Chemical Company, Redwood City, Calif., supplied the S-30 protein adsorbant resin.

Covalent Binding of α-Amylase to Nylon Tubing

α-Amylase was bonded to 1-mm bore-type 6/6 nylon tubing by modification of techniques described by Inman and Hornby (1972). A 3-m coiled section of tubing representing a plane area of 96 cm^2 was perfused at 50° C for 1 hr with a methanolic solution of $CaCl_2$ to remove amorphous nylon. The solution was composed of 18.5 g $CaCl_2$, 18.5 g water, and 63 g of methanol. After rinsing until no Cl^- could be detected, the tube was perfused in a closed loop with 3.65 N HCl at 50° C for 20 hr. Although this acid treatment is more drastic than the one Inman and Hornby (1972) described, it provided considerably more free amino groups without adversely affecting the nylon structure. The tube was thoroughly washed with water until free of Cl^- and then perfused overnight at 25° C with 6% glutaraldehyde in 0.5 M sodium phosphate buffer, pH 7.0. Excess glutaraldehyde was rinsed out with water and a solution containing 2 mg/ml of crystalline α-amylase in 0.1 M phosphate buffer, pH 7.0, was pumped through the tube in a closed loop at 5° C for 40 hr. Alternate washing with water and 0.1 M NaCl removed excess enzyme. Final washing was done by recycling 20 ml of a 1% corn dextrin through the tubing for 75 min.

Bonding α-Amylase to Grafted Cotton Cloth Carriers

Acrylamide was grafted to cotton cloth by γ-irradiation. An 80-square cotton cloth was the substrate in grafting reactions. The cloth before desizing contained starch warp size and had a nitrogen content of 0.02%. Acrylamide monomer was purchased from Eastman Kodak Co., mp 84-86° C. The γ-ray source was a Gammacell 200 well-type ^{60}Co unit with an activity of approximately 8000 curies. During the course of grafting, the dose rate of the unit averaged 1.1 Mrad/hr.

Strips of cotton fabric (32.5 X 5 in.) were desized for graft polymerization by boiling in a dilute NaOH solution at pH 11.6 for 3 hr. After cooling, the strips were neutralized to pH 2.0 with HCl and held for 10 min before washing with Ivory soap. The cloths were then rinsed repeatedly in distilled water and dried at room temperature. The desized cloth did not give a qualitative test for starch with iodine.

Acrylamide was grafted to the cotton fabric strips by simultaneous or mutual irradiation methods. The cloths were soaked in monomer solution, usually at 10% concentration, and deoxygenated by evacuation before irradiation performed at room temperature.

The absorbed dose was generally about 0.055 Mrad. After irradiation, the strips were kept immersed in the monomer solutions for
at least 1/2 hr while post-irradiation polymerization continued.
The cloths were washed repeatedly in distilled water to remove non-
grafted homopolymers and then dried in a vacuum dessicator.

The dried cloths were cut into squares (5 in. X 5 in.) and
reacted with ethylenediamine at 90° C for 5.5 hr to make the amino-
ethyl derivative (Inman and Dintzis, 1969). Excess diamine was
washed out with water, and the derivatized cloth squares were
reacted with 5% glutaraldehyde in 0.5 M sodium phosphate buffer,
pH 7.0, at room temperature for 4 hr followed by washing to remove
free glutaraldehyde. The cloth squares were then immersed in a
2 mg/ml solution of α-amylase in 0.1 M sodium phosphate buffer, pH
7.0, contained in a bottle being continuously turned by a jar
roller. The reaction was allowed to proceed for 16 to 20 hr at
room temperature after which the pieces were washed successively
in running tap water, 0.1 M NaCl, and running distilled water.
Last traces of unbound enzyme were removed by several rinses in
0.1% Lintner starch solution. The cloth squares were considered
to be free of unbound enzyme when on standing the starch supernatant
from the washes failed to increase in reducing sugar value.

Binding α-Amylase to Porous Glass Beads

p-Nitro-benzoylated 500 Å porous glass beads were furnished
by Corning Glass Works, Corning, N.Y. The nitro function was reduced by boiling the glass for 30 min in 1% sodium dithionite
solution. The resulting arylamine glass was washed with water,
followed by acetone, and then air dried. The dry glass was suspended in ice-cold 2 N HCl in a Buchner flask immersed in ice.
One-half gram of solid $NaNO_2$ for each 20 ml of acid was added and
a vacuum applied to remove nitrogen oxides. After 30 min the
diazotized glass was washed on a Buchner funnel with ice-cold water
to remove excess HNO_2 and HCl. Wet cake representing 1 g of dry
glass was added to 25 ml of a 2 mg/ml solution of crystalline α-
amylase in 0.25 M sodium phosphate buffer, pH 8.1. The reaction
mixture was gently agitated at room temperature for 1 hr and was
then thoroughly washed with water.

Alternatively, arylamine glass was reacted with 2.5% glutaraldehyde in 0.1 M sodium phosphate buffer, pH 7.0, under vacuum
for 1 hr at room temperature. Excess glutaraldehyde was removed
by washing with water on a Buchner funnel. The moist glass was
added to a solution of α-amylase in 0.1 M sodium phosphate buffer,
pH 7.0. Fifty milligrams of α-amylase was used per gram of glass.
The mixture was held overnight at 5° C with occasional stirring.
Excess enzyme was removed by washing with copious amounts of water.

Binding α-Amylase to NiO Screen

A 1 X 5 in. 150-mesh nickel screen weighing 1.1 g was oxidized at 700° C in a stream of oxygen. The oxidized screen was refluxed overnight in 70 ml of 10% 3-aminopropyltriethoxysilane in toluene, washed with acetone, and air dried. The amino group was activated by refluxing the screen overnight in 70 ml of 10% thiophosgene in chloroform. After rinsing with chloroform and air drying, it was stirred for 3 hr at 25° C in 30 ml of 0.1 M NaHCO3, pH 9.0, containing 150 mg of crystalline α-amylase. The α-amylase-NiO screen was washed with distilled H2O and stored under water in a refrigerator. The difference in protein content of the original enzyme solution and of the enzyme solution recovered after reaction showed that 2.8 mg of protein was bound to the screen. This procedure is similar to that described by Weetall and Hersch (1970).

Protein Determination

Soluble protein was determined by the method of Lowry et al. (1951). Protein bound to carriers was determined by assaying for arginine according to a method adapted from Messineo (1966) or by measuring tryptophan as described by Gaitonde and Dovey (1970).

Measurement of Enzyme Activity

Soluble and insoluble α-amylase activities were measured by determining reducing values of released oligosaccharides as maltose equivalents. Reducing sugars were determined by the AutoAnalyzer method of Robyt et al. (1972). Degree of starch degradation was arbitrarily defined as the percentage of apparent maltose produced compared to theoretical maltose available from the substrate. One unit of α-amylase is that amount of enzyme required to produce 1 μmol maltose/min at 40° C. In some experiments, the activity was expressed as mg/ml of glucose equivalents using standard AutoAnalyzer procedures for reducing sugar (Technicon Instruments Corp., 1963).

Measurement of Turbidity

Turbidity was measured by light transmittance with either a Gilford spectrophotometer having a 1-cm light path or with a Spectronic 20 colorimeter having optically matched 18-mm test tubes with a light path of 1.4 cm. A Coleman photo-nephelometer Model 7 was selected when greater sensitivity was required. The instrument was calibrated near maximum galvanometer deflection with 0.5% soluble starch. Readings were made by the null-point method.

Turbidity was also measured by APHA Standard Methods for Examination of Water and Waste Water using a standard candle and reporting results as Jackson turbidity units (APHA, 1971).

RESULTS AND DISCUSSION

Duolite S-30-α-Amylase Column

Duolite S-30 is a phenol-formaldehyde resin with protein adsorbant properties. α-Amylase is adsorbed onto the resin and then the protein is crosslinked with glutaraldehyde to keep it from desorbing when in use. The preparation we used had 8.9 mg α-amylase/g resin. A column containing 10 g of active resin was prepared. Substrate was fed to the column by the upflow technique and output was measured as apparent maltose.

The column lost little or no activity during 10 days of operation. At times the rate of conversion appeared to be slowing, but stirring and reforming the bed caused the rate to return to the original value. Apparently some channeling took place even though upward flow was used.

The S-30-α-amylase complex was then stored under water at 5° C for approximately 2 months. It was restarted on a 0.1% soluble starch substrate at a flow rate of 51 ± 2 ml/hr, and no loss of activity occurred during 2 weeks of continuous operation (Figure 1A). An apparent loss of activity after about 6 days is due to physical factors; namely, coating of the particles and possibly channeling. Washing the resin with water followed by 0.1 $\underline{M}$ NaCl and repouring the column restored activity to its initial value. No estimate of half-life can be made at this time for α-amylase on Duolite S-30 since no detectable loss of activity was ever noted.

α-Amylase-Nylon Tube Reactor

Initial experiments involved recycling 20 ml of a 1% solution of a corn dextrin through the coiled tube for various periods of time. It required about 12 recycling runs totaling 25 hr operating time to remove all the unattached enzyme. The tube stabilized with an output of about 0.6 mg reducing sugar per hour as glucose.

The tube was then operated continuously for more than 10 days on 1% dextrin at a flow rate of 8.0 to 9.0 ml/hr. No detectable activity was lost over this time period.

The α-amylase-nylon tube was subsequently left at room temperature for more than 3 months. During this time it was sealed at

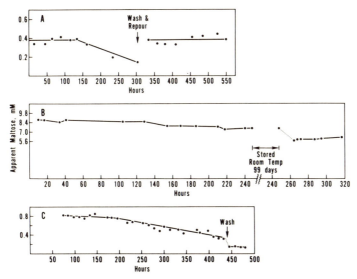

Figure 1. Stability of immobilized α-amylase. A. α-Amylase-
phenolic resin (Duolite S-30) during continuous operation at
50 ml/hr with 0.1% Lintner starch, 40° C, pH 7.0. B. α-Amylase-
nylon coil during continuous operation at 8.5 ml/hr with 1.0%
corn dextrin, 40° C, pH 7.0. C. α-Amylase-polyacrylamide-cotton
square during continuous operation at 50 ml/hr with 0.1% Lintner
starch, 40° C, pH 7.0.

the ends to prevent moisture loss. Upon restarting under similar
conditions using 1% corn dextrin as substrate, the activity had
fallen by 25%. The lower activity level remained constant during
3 days of use (Figure 1B).

α-Amylase Cloth

The cloth square weighing 5 g and containing 5.7% polyacrylamide was suspended in a 250-ml Bellco spinner flask and 0.1%
Lintner starch was fed continuously at a flow rate of approximately
50 ml/hr at 40° C. Figure 1C shows that apparent maltose production was level for the first 200 hr and then declined during the
next 200 hr. The cloth was then removed from the reactor and
washed in running tap water followed by 0.1 M NaCl and then returned
to the reactor. This operation caused greater loss, however, so
that the cloth was performing at only 20% of its peak activity.
The reason for the loss in activity is not known at this time.

Treatment of Starch Solutions

<u>α-Amylase Covalently Linked to Porous Glass Beads</u>. The
enzyme-glass complex had a specific activity of 0.2 μM maltose/mg
glass per min at 40° C.

A 4-1. continuous stirred tank reactor was operated on 1%
soluble starch for 94 hr with a 1.5 ml/min feed rate. One gram of
enzyme-glass complex, contained in a 100-mesh, stainless-steel basket attached to the agitator shaft, was used to convert the substrate. The immobilized α-amylase showed no decrease in activity
over the 94-hr period and about 27% of the starch was converted to
apparent maltose.

A column of α-amylase-glass was prepared and a dialyzed solution of a corn dextrin served as feed. Dialyzing removed 40% of
the lower molecular weight solids and nearly all the reducing
power. On prolonged operation, the column tended to plug but
simple stirring of the bed usually restored the flow rate. Over a
10-day period, the activity decreased by about 50%. The decrease
is mainly attributed to leaching of adsorbed enzyme and to a lesser
extent to solubilization of the glass.

TREATMENT OF PAPER MILL WHITE WATER

White water containing suspended solids from a pilot paper
machine was chosen as a substrate on which to test various forms
of immobilized α-amylase. The effectiveness of the enzyme treatment was measured either by: (1) loss of starch-iodine color,
(2) decrease in turbidity of the white water, or (3) filterable
solids weight.

With α-Amylase-Glass Beads

The first attempts to break the colloidal starch solutions
employed an α-amylase-porous glass complex in a stirred tank reactor. The active glass was contained in a stainless-steel basket
fitted to an agitator shaft. The rotation of the basket through
white water was sufficient to get enzymatic breakdown of the
starch. A solution of all white-water ingredients except starch
was used as a control. After treatment with the glass-enzyme
complex, samples were allowed to stand for 24 hr before measuring
turbidities with a nephelometer. No alum was used in this experiment. A brief enzyme treatment is sufficient to settle the solids
nearly as well as when no starch is present (Table I). Effectiveness of the glass-enzyme complex was also measured by determination
of filterable solids in treated and untreated white water (Table II).
Solutions from the various treatments were filtered through a

Table I. Reduction of Turbidity of White
Water Containing Unmodified Wheat Starch with
α-Amylase-Porous Glass Complex

Trial No.	Time of treatment, min	Turbidity[a] starch in white water	
		0.01%	0.02%
1	5	2.50	1.82
2	5	2.45	2.23
3	10	2.00	2.45
4	No treatment	4.72	>13.0

Conditions: 1 g active glass, 2 1. white water
40° C.

[a] Nephelometer readings taken 24 hr after
treatment. Control white water (no starch)
gave turbidity reading of 1.40 after standing
24 hr.

Table II. Properties of White Water[a] Treated with
α-Amylase-Glass Complex

Treatment	Starch-iodine color, O.D.[b]	Turbidity, O.D.[b]	Filtrate solids, mg/100 ml
Control	0.697	0.045	240
100 mg Glass-enzyme[c]	0.523	0.017	15
300 mg Glass-enzyme[c]	0.286	0.003	13

[a] White water contained 0.015% Lintner starch.

[b] O.D. read at 590 nm, 1-cm light path after 18 hr
settling.

[c] Reaction time--10 min, 40° C.

coarse sintered glass filter, and solids were determined on the
filtrates. For this experiment, the glass-α-amylase complex was
added directly to the white water. When 100 mg of bound enzyme
was used, there was little loss of iodine color. Apparently, it
is not necessary to hydrolyze the starch to the achroic point in
order to reduce significantly the suspended solids.

With a NiO-α-Amylase Screen

A 1-g nickel oxide screen to which 2.8 mg α-amylase had been
bound was tested for its ability to degrade starch in white water.
The screen was rolled into the form of a cylinder about 1/2 X 1 in.
and dipped into 20 ml of stirred white water that contained 0.02%
starch. At definite time intervals, the screen was removed and
samples were taken for measurement of starch-iodine color. Eight
tests were conducted with the screen. Although the α-amylase-active
screen is able to degrade the starch, loss of activity was quite
pronounced over the eight trials. The magnitude of enzyme loss is
illustrated by Figure 2. The loss of activity was probably due to
flaking off of the oxide from the screen.

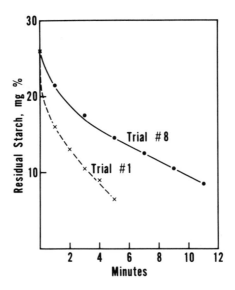

Figure 2. Stability of α-amylase-NiO screen during eight trials
in 20 ml of white water containing 0.02% starch at room
temperature, pH 7.3.

With an α-Amylase-Nylon Tube

The α-amylase-active nylon tube, which was used in experiments described previously, also served to treat stock white water prepared as described under Materials and Methods. The white water, stirred continuously to keep the solids suspended, was perfused through the tube with the aid of a peristaltic pump. Residence time of the white water in the tube varied from 15 to 30 min. Table III contains some results with Penford gum, a hydroxyethyl starch.

Table III. Treatment of White Water[a] with an α-Amylase-Nylon Tube

Hydroxyethylated starch, %	Contact time, min	Turbidity			
		Initial		Settled 1 hr	
		O.D.[b]	JTU[c]	O.D.	JTU
0.01	0	1.00	900	0.32	220
0.01	15[d]	0.95 ± 0.10	730 ± 150	0.18 ± 0.02	130 ± 15
0.02	0	1.00	900	0.45	310
0.02	30[d]	0.73 ± 0.03	510 ± 25	0.33 ± 0.01	220 ± 15

[a] Stock white water with hydroxyethylated starch added.

[b] O.D., 650 nm, 1.4-cm light path.

[c] JTU = Jackson turbidity units.

[d] Average of three samples collected over a 24-hr period.

Samples were taken at periodic intervals and mixed to obtain initial turbidity values. After 1 hr, the turbidity was read again on the unmixed samples and compared to untreated white water similarly handled. Even during a 15-min contact time, the starch broke down sufficiently to allow more rapid settling than in the control. With a 0.02% hydroxyethyl starch solution, a longer contact time is required to get a similar settling pattern. As shown in Table IV, the rate of settling of α-amylase-treated white water is somewhat enhanced if alum is added after enzyme treatment. Untreated white water showed a decrease in turbidity from an O.D. of about 1.0 to O.D. ~0.25 after 3 hr settling. Alum (40 ppm) did not affect the rate of settling. When the white water was treated with α-amylase by passage through the nylon tube-enzyme reactor, the initial turbidity after remixing was around 0.75 either with

Table IV. Settling Characteristics of White Water[a] Treated with
an α-Amylase-Nylon Tube and 40 ppm Alum

| Settling time, hr | Untreated | | | | α-Amylase treated | | | |
| | No alum | | Alum | | No alum | | Alum | |
	O.D.[b]	JTU	O.D.	JTU	O.D.	JTU	O.D.	JTU
0	1.05	>900	1.05	>900	0.73	500	0.72	500
1	0.43	290	0.41	280	0.18	120	0.13	80
3	0.25	160	0.29	190	0.10	60	0.07	35

[a] Stock white water plus 0.02% hydroxyethyl starch in solution.

[b] 650 nm, 1.4-cm light path.

or without alum. After 3 hr settling, the tubes containing alum
had turbidity readings of 0.07, which would probably be considered
satisfactory for reuse in a plant or for discharge as wastewater.

The microenvironment of an immobilized enzyme particle is
known to have a significant effect on activity in at least three
ways: charge-charge interaction between substrate and particle
(Hornby et al., 1968); diffusion of substrate to the surface of
the particle (Goldman et al., 1971); and steric effects (Goldman
et al., 1971).

Unmodified starch is essentially without charge so that
charge-charge relationships will be of little significance in the
immobilized α-amylase reaction. The paper industry uses a variety
of modified starches, among them cationic starch. Degradation of
this positively charged starch would be inhibited by α-amylase
bound to a positively charged carrier. A negatively charged
carrier could be expected to accelerate the hydrolytic reaction
due to attraction of the substrate to the enzyme-carrier complex.
Obviously the choice of carrier will be influenced by the type of
starch to be degraded.

Diffusion of substrate to the surface of the enzyme carrier
may also limit the rate. The immobilized enzyme particle can be
presumed to be surrounded by a near stagnant film of solvent,
sometimes referred to as the Nernst diffusion layer, through which
the substrate must diffuse to reach the enzyme. The rate of transfer through the film is directly proportional to the substrate
concentration but inversely porportional to its molecular weight

and to the film thickness. Film thickness is decreased by increasing the stirring rate in stirred tank reactors or by increasing flow rates in column and fluid bed reactors (Goldman et al., 1971; Rovito and Kittrell, 1973). The α-amylase-Duolite S-30 system was used to study the effects of diffusion. Starch concentrations ranging from 2.5 to 12.5 times greater than the Km value of the free enzyme were perfused at 8.5 ml/hr through the column of α-amylase-Duolite S-30 described previously. The Km value was presumed to be about 0.04% starch (Fischer and Stein, 1960). The amount of apparent maltose produced was directly proportional to starch concentration in the range of 0.1 to 0.5%, indicating that diffusion does play a role in the immobilized α-amylase reaction. The role of diffusion was further confirmed by measuring the effect of flow rate on the amount of apparent maltose formed per minute. As mentioned above, the velocity of substrate flow through a column of immobilized enzyme particles will determine the thickness of the solvent film surrounding the particle (Hornby et al., 1968). Increasing the flow rate of a 0.1% starch solution from 1.7 ml/min to 3.0 ml/min increased apparent maltose production from 0.136 μmol/ml/min to 0.225 μmol/ml/min. Further increases in flow rate failed to increase the rate of apparent maltose production, indicating that the solvent film thickness had reached a minimum value.

Steric effects are also known to influence immobilized enzyme reactions (Goldman et al., 1971). The composition of a starch solution is complex. Starch is a mixture of linear (amylose) and highly branched (amylopectin) glucans. Both fractions show a rather wide range of molecular sizes. In addition, the large molecules interact to form agglomerates. These particles may be sterically hindered from approaching amylase attached to an insoluble carrier. Evidence of such an effect was obtained using the nylon tube reactor. A dilute starch solution (0.05%) was pumped through the tube at 8.5 ml/hr, which corresponds to a residence time of 50 min. Positive starch-iodine color remained although 0.08 mg/ml of apparent glucose was formed, representing the rupture of 16% of the available bonds in the starch molecule. This degree of hydrolysis is adequate to degrade starch beyond the iodine achroic point when soluble α-amylase is employed. If the resistant starch is exposed for a longer time to the insoluble α-amylase, it will eventually be degraded beyond the achroic point. These results indicate that there is a degree of steric inhibition of starch hydrolysis by the α-amylase-Duolite S-30 complex. Ledingham and Hornby (1969) noticed a greater degree of multiple chain attack on starch by immobilized α-amylase compared to soluble α-amylase. They attributed the phenomenon to steric effects. Our results seem to support their conclusion.

The aspects of starch degradation discussed above, i.e., charge-charge interaction, diffusional barriers, and steric considerations, do not preclude use of immobilized α-amylase for treatment of paper mill waste streams. The experiments described in this report have demonstrated that suspended solids in white water readily settle after a brief treatment with immobilized α-amylase.

Evidently, the colloidal nature of the starch solution has been altered to the extent that residual iodine staining material does not prevent flocculation of suspended solids.

CONCLUSIONS

The results indicate that α-amylase attached to nylon, Duolite S-30, or a radiation-grafted copolymer of acrylamide-cotton has the capability of degrading starches in wastewater streams and thereby of significantly improving primary clarification. A Duolite S-30-α-amylase column has been operated on soluble starch for 950 hr without measurable loss of activity. An α-amylase-nylon tube was operated approximately 500 hr on various substrates, including white water, with only slight loss of activity. White water operation was about 120 hr. Both nylon tubing and cotton cloth readily lend themselves to reactor design and should be practical for treatment of white water. Duolite S-30, operated as a column, should be practical for white-water treatment if the water is freed of large particles. Both α-amylase-active porous glass and a nickel screen were too short-lived to be of commercial value.

Schwanke and Davis (1973) have shown that soluble amylase enhances the ability to flocculate paper mill effluents. Because a soluble enzyme must be destroyed before clarified water can be used in paper making, an immobilized enzyme eliminates this drawback.

SUMMARY

White water from paper mills is typical of a colloidal starch waste stream containing suspended solids. Laboratory-scale experiments demonstrated that immobilized α-amylase can degrade starch in paper mill-type effluents. When bound to either nylon, protein-adsorbent resin, or polyacrylamide-cotton cloth supports, α-amylase was stable for prolonged periods of continuous use. Settling rates of white-water solids were improved by treatment with immobilized amylase followed by addition of alum.

LITERATURE CITED

Fischer, E. H., Stein, E. A., in "The Enzymes," Boyer, P. D., Lardy, H., Myrback, K., Ed., Vol. 4, Academic Press, New York, Chap. 19, 1960.

Gaitonde, M. K., Dovey, T., Biochem. J. 117, 907 (1970).

Goldman, R., Goldstein, L., Katchalski, E., in "Biochemical Aspects of Reactions on Solid Supports," Stark, G. R., Ed., Academic Press, New York, 1971, p 62.

Hornby, W. E., Lilly, M. D., Crook, E. M., Biochem. J. 107, 669 (1968).

Inman, D. J., Hornby, W. E., Biochem. J. 129, 255 (1972).

Inman, J. K., Dintzis, H. M., Biochemistry 8, 4074 (1969).

Ledingham, W. M., Hornby, W. E., FEBS Lett. 5, 118 (1969).

Lowry, O. H., Rosebrough, N. J., Farr, A. L., Randall, R. J., J. Biol. Chem. 193, 265 (1951).

Messineo, L., Arch. Biochem. Biophys. 117, 534 (1966).

Olson, A. C., Stanley, W. L., J. Agr. Food Chem. 21, 440 (1973).

Robyt, J. F., Ackerman, Rosalie J., Keng, J. C., Anal. Biochem. 45, 517 (1972).

Rovito, B. J., Kittrell, J. R., Biotechnol. Bioeng. 15, 143 (1973).

Schwanke, P. A., Davis, W. S., Tappi 56, 93 (1973).

Standard Methods for Examination of Water and Waste Water, 13th ed. Amer. Pub. Health Ass., Inc., New York, N.Y. (1971).

Technicon Instruments Corp., Technicon AutoAnalyzer, Methodology N-2A, Chauncey, New York (1963).

Weetall, H. H., Hersch, L. S., Biochim. Biophys. Acta 206, 54 (1970).

IMMOBILIZED GLUCOSE OXIDASE AND CATALASE IN CONTROLLED PORE TITANIA

Ralph A. Messing

Corning Glass Works

Corning, New York 14830

The immobilization of glucose oxidase and catalase by adsorption within the pores of controlled-pore titania has indicated that catalase acts as a stabilizer for glucose oxidase in this material. Flow rates effect the apparent activity of the immobilized enzyme system. Carrier parameters were varied to obtain optimum loading and stability information.

Initial adsorption studies of glucose oxidase in the pores of 68 Å porous glass indicated that little or no glucose oxidase was bound to the internal surface of the pores (Messing, 1970). Weetall (1970) demonstrated that glucose oxidase could be silane coupled to the internal surface of 735 Å porous glass. This early work with inorganic carriers indicated that additional studies were required to optimize the immobilization of this enzyme with respect to pore size. Since glucose oxidase and catalase may be considered synergistic, the possibility of a simultaneous immobilization was quite intriguing.

Experimental

The glucose oxidase-catalase, a standardized preparation (which is sold under the name of DeeO liquid) purchased from the Marshall Division of Miles Laboratories, Inc., contained 750 glucose oxidase units (GOU) per ml and 225 EU of catalase per ml. The differential conductivity meter was a Wescan Instruments Inc. Model 211. Conductivity flow cells, Model 219-020, having a cell constant K = 80 was obtained from Wescan Instruments

Inc. A four-channel peristaltic pump was fitted with
1/16" I.D. Tygon® tubing in two channels. The columns
used for these studies were Corning Glass Works Code
442802 and 442804 Teflon stopper, straight-bore stop-
cocks, fitted with fiber glass retainers just above the
stopcock fittings. The O.D.'s of these columns were 9
and 10 mm, respectively. The column lengths were approximately 11 cm.

The column conductivity assays were performed at 22°C
employing 6% glucose in 0.0045% hydrogen peroxide in a
manner similar to that described by Messing (1973), except that a 125-ml Erlenmeyer flask containing 100 ml of
glucose substrate solution with a magnetic stirring bar
was utilized for the reaction mixture. This flask was
mounted on a magnetic stirrer and stirring was commenced.
The inlet and outlet tubes were inserted below the surface of the substrate solution and circulation was initiated by turning on the pump. The meter and recorder
were then balanced and, after a stable base line was
achieved, the outlet tubing from the flow cell was in-

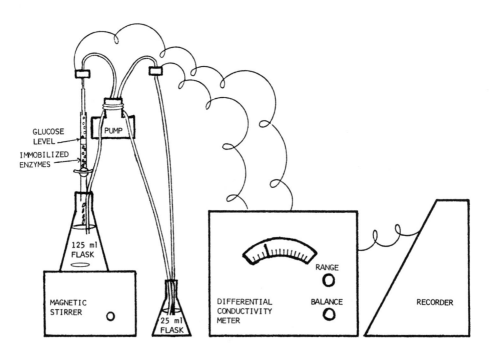

Figure 1. Differential conductivity equipment with immobilized
enzyme column.

serted into the top of the column containing the immobilized enzyme. The stopcock was adjusted to maintain
a 3/4" head above the immobilized enzyme as the substrate
solution circulated through the column and back into the
reaction flask. This circulation was maintained throughout the assay. (See Fig. 1).

The carriers used for these studies were selected
from a spectrum of porous inorganic oxides tailored and
developed by the author for the specific purpose of immobilizing biologically active macromolecules. The physical
properties of the carriers are summarized below and the
carriers, hereafter, are specifically referred to in terms
of their average pore diameter and their chemical composition.

	Al_2O_3	44% TiO_2 56% Al_2O_3	TiO_2	TiO_2	TiO_2	TiO_2
Ave Pore Diam (Å)	175	220	350	420	820	855
Min Pore Diam (Å)	140	140	220	300	760	725
Max Pore Diam (Å)	220	300	400	590	875	985
Pore Volume (cm^3/g)	0.6	0.5	0.45	0.4	0.2	0.22
Surface area (m^2/g)	100	75	48	35	7	9
Particle mesh size	25-60	25-60	25-60	30-80	25-80	25-80

The immobilization of the enzyme was performed
either in a 10-ml cylinder or a 25-ml Erlenmeyer flask.
A volume of dialyzed glucose oxidase-catalase solution
(containing either 2400 or 4500 GOU) was added to 300 mg
of carrier which had been preconditioned with 0.5 M
$NaHCO_3$. The vessel was then placed in a shaking water
bath at 35°C and reacted with shaking for at least two
hours and 20 minutes. The vessel was then removed from
the bath and the adsorption and diffusion was allowed to
continue overnight (approximately 15 hours) at room temperature without shaking. The enzyme solution was de-
canted and the immobilized enzyme system was washed with
9 ml volumes of distilled water, 0.5 M NaCl, 0.2 M acetate
buffer pH 6.1 and finally distilled water. The immobilized enzyme system was then transferred with distilled
water to the stopcock columns previously described. The
immobilized enzymes were stored between assays at room
temperature in the columns filled with distilled water
and stoppered with cork stoppers.

The results presented in this paper were obtained
from the initial slopes of the differential conductivity
recordings which were then multiplied by the cell constant
80, corrected for dilution by multiplying by four (100 ml
of substrate was used in place of the 25 ml calibration

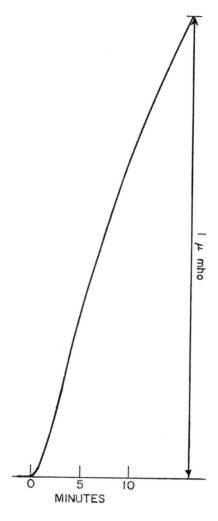

Figure 2. Differential conductivity recording of immobilized
enzyme.

quantity), correlated to GOU by dividing by 2.65 and
finally reduced to a per-gram base by multiplying by
3.33. A typical recording of the immobilized enzyme
reaction may be seen in Fig. 2. It will be noted that

in this recording there is no contribution to conductivity by the enzyme itself during the initial minute as
reported for the free enzyme (Messing, 1973).

Since the procedures were not identical with respect
to enzyme charge and time of exposure at 35°C during the
preparation of immobilized enzymes, these will be reported for each of the following studies.

The immobilized enzyme for the flow rate and peroxide studies was prepared by exposing 2400 GOU to 350 Å
titania at 35°C for two hours and 20 minutes.

Flow Rate (ml per hour)	Apparent Activity (GOU per gram)
125	30.2
182	46.3
235	52.4
280	60.3
390	66.4

It is not at all surprising to note that increases
in flow rate result in increases in the apparent activity.
Increasing flow rates of the substrate solution not only
removes the reaction products from the pores but renews
the concentration of glucose at the surface of the carrier, thus increasing the diffusion of the glucose. Flow
rate increases above 350 ml per hour have little or no
effect upon increasing the apparent activity.

It was previously noted (Messing, 1973) that glucose substrates devoid of hydrogen peroxide could not be
utilized to attain reproducible results. By increasing
the hydrogen peroxide concentration from 0.0045% to
0.0075%, a factor of 1.67, we increased the apparent activity from 35.9 to 48.3 GOU per gram, a factor of 1.34
at a flow rate of 145 ml per hour. Although we have not
reached the optimum concentration of peroxide in these
studies, it is clear that increasing concentrations of
peroxide facilitates the glucose oxidase activity. Attempts to increase the concentration above 0.0075% H_2O_2
resulted in gas entrapment within the column and a reduced flow rate.

The alumina and 350 Å titania immobilized enzymes
were prepared at the same time and exposed to the same
quantities of enzyme under the same conditions. The
titania-bound enzyme is the same preparation reported
in the previous studies. The activity determinations
were performed at a flow rate of 145 ml per hour.

	Apparent Activity (GOU/g)	
Assay Day	175 Å Alumina	350 Å Titania
1	12.9	20.5
1	18.7	36.2
4	11.7	36.2
5	10.1	36.2
6	8.5	36.2
7	7.0	36.2
38	Results too low	32.5
	to measure	
59	---	36.2
83	---	35.9
103	---	35.5
136	---	32.5
165	---	29.6

The immobilized enzyme system used to study 220 Å titania-alumina, 420 Å, 820 Å and 855 Å titania carriers were prepared by exposing the carrier to 4500 GOU over a five-hour period at 35°C. The activity determination was performed at a flow rate of 390 ml per hour.

		Apparent Activity (GOU/g)		
Assay Day	220 Å	420 Å	820 Å	855 Å
0	23.7	56.4	43.9	37.8
3	--	77.7	43.9	39.7
4	11.7	--	--	--
8	--	84.5	46.2	40.3
12	9.25	--	--	--
13	--	84.5	46.2	44.3
18	6.83	--	--	--
42	--	80.5	40.3	42.0

Result and Discussion

Although the immobilized enzyme systems were prepared by exposing the carriers to two different quantities of enzymes and two different time intervals, several points become apparent when the accumulated data is compared at flow rates of 390 ml per hour. Immobilized enzyme preparations prepared with 175 Å alumina and 220 Å titania-alumina proved to be unstable. On the other hand, a very stable glucose oxidase was achieved with 350 Å titania and stable preparations were achieved with 420 Å, 820 Å and 855 Å titania. If we examine these results with relation to the major dimensions of the unit cell of these enzymes, a clearer picture evolves. Catalase has a major axis of 183 Å. This molecule spinning in solution probably occupies a sphere having a diameter of twice the major axis, or 366 Å. This spinning molecule could not possibly enter a pore having dimensions

significantly less than a 366 Å diameter. Thus, all of
the catalase would be immobilized at the mouth of the
pore rather than entering the internal structure. The
same argument could be applied to the glucose oxidase
molecule with its major dimension of 84 Å. The sphere
occupied by the glucose oxidase would have a diameter
of approximately 168 Å. It is apparent, therefore, that
neither one of these enzymes could be immobilized within
pores significantly below 168 Å. Elevations in peroxide
concentrations increase the apparent activity of the
glucose oxidase; thus it can be implied that catalase
recycles the peroxide produced by the glucose-glucose
oxidase reaction internal to the porous materials. If
catalase is excluded from the pores, then the apparent
activity of the glucose oxidase should be very low. This
was true in both the case of 175 Å alumina and 220 Å
titania alumina. It has been noted, in addition, that
glucose oxidase free of catalase is very unstable in
solutions containing small quantities of hydrogen peroxide. The 350 Å titania, which approaches the spin diameter of the catalase, is the smallest-pore material that
demonstrates a stable glucose oxidase preparation. It
is probable that the accummulated peroxide within the
pores containing glucose oxidase, but not catalase, causes the oxidative degradation of glucose oxidase. Thus,
the catalase exerts a protective effect upon the glucose oxidase.

Although no definitive conclusion can be drawn, it
would appear that the most stable immobilized glucose
oxidase-catalase system was achieved in the 350 Å titania carrier. This carrier contained the smallest pores
that would accommodate both the glucose oxidase and catalase. From this it could be conjectured that smaller
pores exert greater protective effects upon the enzyme
activity. If one considers the turbulent outer environment existing in a column that flows at approximately
145 ml per hour, it is not surprising to see that the
smaller the pore, the less the turbulence within the pore
and, thus, less destruction of the tertiary enzyme structure would be experienced.

It would appear that the highest loading with respect to apparent glucose oxidase activity was achieved
in 420 Å titania. This should not be considered con-
clusive, since the 350 Å titania was loaded at a lower
charge level for a shorter period of time; however, a
valid comparison exists between the larger-pore carriers.

Acknowledgment

I am truly grateful for David L. Eaton's assistance in evaluating and preparing quanties of porous inorganic oxides. In addition, Mr. Eaton has contributed to the increased utility of these materials.

Literature Cited

Messing, R.A., "Enzymologia", 39, 12 (1970).
Messing, R.A., Biotech. Bioeng., "Assay of Glucose Oxidase by Differential Conductivity" (1973). Accepted for publication.
Weetall, H.H., Biochimica et Biophysica Acta 212, 1 (1970).

COLLAGEN AS A CARRIER FOR ENZYMES: MATERIALS SCIENCE AND PROCESS ENGINEERING ASPECTS OF ENZYME ENGINEERING

Fred R. Bernath and Wolf R. Vieth

Department of Chemical and Biochemical Engineering

Rutgers University, New Brunswick, N.J.

INTRODUCTION

During the last fifty years a wealth of information has been accumulated concerning the structure and function of enzymes. Although we are probably still far from achieving a total under-standing of the way enzymes act on a molecular level, we certainly have sufficient knowledge to begin designing large scale processes that capitalize on the unique properties of these biological cata-lysts. This is especially true for the relatively simple extra-cellular hydrolytic enzymes.

Actually, we have probably had the capability of developing enzyme applications for a number of years. The primary emphasis in enzymology, however, has been placed on what has sometimes seemed to be a rather narrow track of basic research while other important aspects have in many cases been neglected. Even studies of immobilized enzymes, which were originally stimulated by prac-ticable objectives, have been relatively basic in nature rather than applications-oriented. A large number of immobilized enzyme systems have been developed and studied with relatively little regard for considering criteria for their generality and/or tech-nological potential. In particular, most research has dealt with the comparison of kinetics and mechanistics of soluble and immo-bilized forms of enzymes, determination of physical changes in the enzyme molecule in its microenvironment and other similar studies. This work is certainly valuable but unfortunately has often been conducted exclusively at the expense of such considerations as basic materials science, reactor design and scale-up, with ef-fective treatment of mass transfer effects, feasibility and cost analyses, and other important aspects.

157

Within the last five years or so a new area of endeavor has been emerging to deal with the applied research and engineering aspects of enzyme technology. This area, which some have called enzyme engineering, is currently being defined by the engineers, chemists, biochemists, microbiologists, food technologists, and others who are working to develop applications in industry, medicine and the environment. There currently appear to be two major thrusts or directions in the field. The first is a concentration on the design, operation and scale-up of immediate applications for the simpler well-defined enzymes such as the hydrolases. The second is a continuing research effort aimed at the more complex enzyme systems such as the synthetases. The former course will demonstrate the value and feasibility of enzyme processes, stimulate the interest of potential users and provide a foundation for the more complex developments that are sure to follow. The latter course will hopefully provide these developments. As the principles of enzyme engineering are developed and applied to specific problems, we should achieve a more reasonable balance between basic and applied research in enzymology. This balance should contribute significantly to the realization of the great potential that has been forecast for enzymes.

In this paper we shall discuss the development of a novel process for the general utilization of enzymes in industrial, medical and environmental applications. We shall discuss specific applications of the process, but in so doing shall attempt to outline some of the important principles of enzyme engineering, especially the materials science and process engineering aspects. In addition to the presentation of some novel ideas and new experimental data, we would hope that our work may serve as a model for the rational development of practical enzyme processes.

MATERIALS AND METHODS

Collagen membranes were prepared in our laboratory from both cattlehide and cow tendon. The method described in U.S. Patent No. 2,920,000 (Hochstadt, Park and Lieberman 1960) was followed to prepare 1% (w/w) cow tendon collagen dispersion in a mixed solvent of water and methanol (1:1 by volume). Ground cattlehide (1.3:1 water to hide) was obtained from the U.S. Department of Agriculture Research Service. Three different methods were used to prepare insoluble collagen enzyme complexes from the above dispersions. These methods are illustrated in Figure 2 of the following section.

The assay system consisted of a continuous recirculation loop including, in series, a modular reactor (see next section), a substrate reservoir and an analytical compartment. Adjustment of the ratio of reactor volume to reservoir volume allowed operation in a number of modes between the differential and integral reactor modes. This system allowed a rapid and simple evaluation of the activity and stability of the immobilized enzyme.

Lactase activity was followed by measuring glucose production by the glucostat method. Lysozyme was assayed by observing the decrease in turbidity of a cell wall suspension at 450 nm. Glucose isomerase activity was followed by measuring fructose production by the color developing reaction (420 nm) between fructose and thiobarbituric acid. Urease and 1-asparaginase activities were measured by following the production of ammonia by a potentiometric technique. Invertase activity was followed polarimetrically, glucose oxidase by the decrease in dissolved oxygen measured by an oxygen electrode and catalase by the decrease in hydrogen peroxide measured by iodometric titration.

Lysozyme from hen egg white and lactase from E. coli were obtained from Worthington Biochemical Corporation, Freehold, N.J. Invertase from yeast and catalase from beef liver were obtained from Mann Research Lab, New York, N.Y. Glucose oxidase from Aspergillus niger was obtained from Miles Lab, Kankakee, Ill.; jack bean urease was obtained from Nutritional Biochemicals Corporation, Cleveland, Ohio; and 1-asparaginase from E. coli was the generous gift of Merck, Sharp and Dohme of West Point, Pa. Glucose isomerase was produced in our laboratory by fermentation of Streptomyces venzuelae, NRRL B-3559 and NRRL B-5333.

RESULTS AND DISCUSSION

Background

Enzyme engineers are involved in a number of steps in the development of a commercial enzyme process. These operations which are shown in Figure 1 include production and purification of enzymes, utilization of the catalyst and separation and purification of products. The industrial use of enzymes has been severely limited by difficulties that have arisen in the first two areas. The problems include a relatively high cost of the enzyme due to isolation and purification techniques and the inability to reuse or regenerate the catalyst. Both factors have made many potential enzyme processes economically prohibitive.

Attempts have been made to lower the high initial cost of some enzymes by developing continuous, high capacity, high resolution purification techniques. The problem of regenerability has been approached from the standpoint of immobilized enzyme systems. We decided to concentrate our efforts on the latter approach in our attempts to design practical enzyme processes

for areas where immediate socioeconomic need exists. We made this
decision based on our belief that reactor performance, measured
by such characteristics as apparent activity, stability and re-
usability, is the single most important parameter in determining
the economic feasibility of a process. An ideal development would
be a simple, inexpensive reactor and carrier system that is limited
only by the cost of the enzyme being used. If such a system were
available a shift in emphasis could then be made to enzyme purification studies for specific processes where further improvements
are necessary.

Choice of Collagen as Carrier

We directed our first efforts toward choosing a suitable
carrier. An investigation of existing carriers and their corresponding advantages and disadvantages provided an appreciation of
the properties of an ideal enzyme matrix. Since each existing
carrier appeared to have some limitations, we investigated other
available materials which might incorporate as many of the properties of an ideal carrier as possible. The carrier material must
be inexpensive. The immobilization technique must be simple and
be conducted under mild conditions. The carrier must bind a large
amount of active enzyme and maintain stability over long periods
of use and storage. It must also have good mechanical properties,
offer low resistance to substrate diffusion and have a versatile
chemical nature.

An analysis of the above properties suggested that an ideal
carrier might be a film-forming protein. We chose to investigate
collagen for a number of reasons. First of all, it is the most
abundant protein in the higher vertebrates, comprising 30% or more
of the body's total protein (Lehninger 1970). Secondly, its natural functions, some of which are listed in Table 1, give an indication of its strength and versatility (Woessner 1968). Finally,
it has been demonstrated that collagen can be isolated from a number of biological sources, reconstituted into its native microfi-

Table 1

Some In Vivo Functions of Collagen:

1. Impart strength to blood vessels
2. Hold bone crystals together
3. Enable muscles to pull on bones
4. Hold individual cells together in tissue

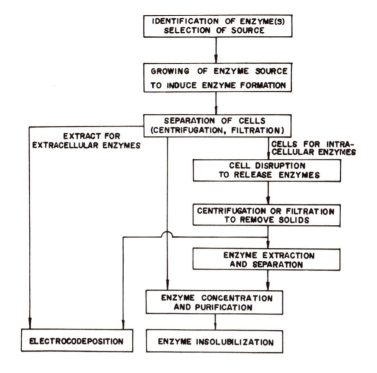

Figure 1. Enzyme Production, Purification and Utilization

brillar configuration and cast into various forms such as membranes, tubes and other configurations (Rubin and Stenzel 1969).
The reconstituted material maintains the same structural and
strengthening properties of collagen and is also subject to a number of useful chemical modifications as shown in Table 2.

Table 2

Possible Modifications of Reconstituted Collagen

1. Crosslinking to produce stronger, tighter structure
2. Treatment with proteolytic enzymes to eliminate anti-
    genic activity
3. Blocking of carboxyl or amino groups to alter surface
    charge
4. Embedding with heparin to minimize clotting of blood on
    surface

Our work over the last four years has demonstrated that reconstituted collagen is indeed an excellent carrier material.
Table 3 lists some of the advantages that we have observed. It
is readily available from a number of sources, especially cattle
and fish, and is quite inexpensive. The immobilization technique
is simple, requiring only collagen, the enzyme, buffer solutions
and in some cases a bifunctional crosslinking agent. In addition,
immobilization takes place at room temperature and in an aqueous
environment. Since collagen is a protein it has a large number
of polar and nonpolar amino acid residues which provide sites for
strong cooperative noncovalent interactions between enzyme and
carrier. The hydrophilic, proteinaceous nature of the carrier
also tends to have a stabilizing effect on the bound enzyme. Collagen sorbs water at levels of 100% of its dry weight at neutral
pH and up to 500% of its dry weight at pH 2 (Bowes and Kenten
1948). This high swelling capacity provides an aqueous environment for the bound enzyme and reduces internal diffusional resistances. Finally, as mentioned above, its fibrous nature provides
a strong material that can be cast into a number of forms, and its
chemical nature enables a wide variety of characteristics via controlled chemical modification.

To date, we have immobilized almost twenty different enzymes
and two strains of whole cells. These are listed in Table 4. In
all cases immobilization was achieved by one of the three methods
described in Figure 2. Initially, we used collagen derived from
cow tendon but have since switched to hide collagen which produces
membranes with superior mechanical properties. Our first complexes
were fabricated by the impregnation method. Since this method
requires about five days, we developed an alternative method of
electrocodeposition which takes advantage of the electrophoretic
properties of collagen. This method requires approximately seven
hours and dispenses with the need for an annealing step. Finally,

Table 3

Advantages of Collagen as Enzyme Carrier

1. Inexpensive
2. Simple immobilization technique; mild conditions
3. High density of reactive groups
4. Hydrophilic proteinaceous nature
5. High swelling capacity in aqueous solution
6. Fibrous nature
7. Variety of characteristics possible via controlled
    chemical modification

Table 4

Enzymes and Whole Cells Immobilized on Collagen

Enzymes

α and β -amylase	glucose isomerase	lactase	rennin
l-asparaginase	glucose oxidase	lysozyme	tyrosinase
catalase	hesperidinase	papain	urease
dextranase	invertase	pectinase	penicillin amidase

Whole Cells

Streptomyces venzuelae (glucose isomerase)
Corynebacterium simplex (steroid conversion)

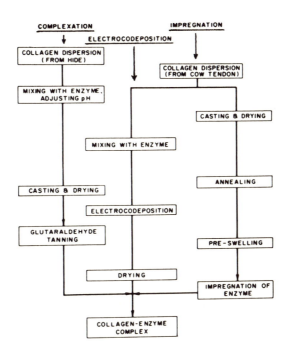

Figure 2. Preparation of Collagen-Enzyme Complexes

we have recently shown that most enzymes can be immobilized by
macromolecular complexation which involves simply mixing an enzyme with collagen dispersion, casting and drying the membrane,
and tanning with glutaraldehyde to improve mechanical strength.
This method, which requires about eight hours, provides a simple,
inexpensive and practical way of preparing immobilized enzymes.
The three methods together offer a valuable flexibility that appears capable of accommodating a large number of different enzymes
and whole cells.

The fact that collagen can successfully bind whole cells has
some very valuable implications. The cost of processes utilizing
intracellular enzymes could be greatly decreased by using whole
cells and dispensing with expensive isolation and purification
procedures. Also, for processes requiring coenzymes an immobilized
whole cell system may be superior to attempts to immobilize coenzymes or provide them in dissolved form with substrate feed
streams.

Properties of Collagen - Enzyme Complexes

Each of the methods discussed above produce collagen - enzyme
complexes with similar properties, indicating that the binding
mechanism is a general one. In all cases a certain amount of
weakly bound enzyme is washed off during initial use before a

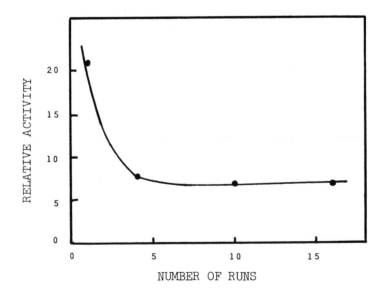

Figure 3.   Stable Limit of Activity for Collagen - Immobilized
            Lactase

stable activity limit is achieved. Figure 3 demonstrates that a
stable limit of activity for a collagen - lactase membrane made
by impregnation is reached after four consecutive runs (Eskamani
1972). This stable limit represents approximately 30% of the
membrane's initial activity and corresponds to about 150 IU/gram
complex. This activity level was maintained through 16 consecutive runs and after four months of intermittent storage and use.
Improvements in the immobilization procedure have produced collagen - lactase complexes with specific activities of 670 I.U./gram
complex at the stable limit. Figure 4 demonstrates similar behavior for a collagen - lysozyme complex made by impregnation
(Venkatasubramanian, Vieth and Wang 1972). A stable activity level of approximately 3,000 lysozyme units/gram complex was maintained over a period of five months of intermittent use and storage
at 4°C. This membrane also retained its activity after two years
of cold storage. Examples of stable limits for other enzymes include glucose oxidase (electrocodeposition) 110 I.U./gram complex,
invertase (impregnation) 300 I.U./gram, and urease (impregnation)
270 I.U./gram. In all cases where glutaraldehyde was not used in
the immobilization procedure stable limits representing 15 - 35%
of the complex's initial (overloaded) activity was observed.
These results indicate the high levels of activity and stability
of collagen - enzyme complexes.

     The use of a crosslinking agent such as glutaraldehyde di-
rectly after immobilization increases the loading capacity of collagen membranes. Figure 5 shows that an 1-asparaginase complex
made by macromolecular complexation retains 67% of its initial
activity or 200 I.U./gram complex through six consecutive runs

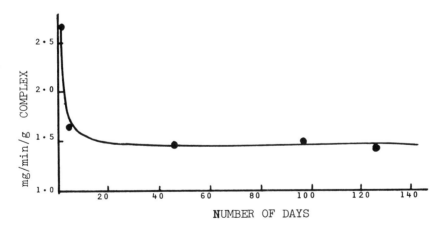

Figure 4. Stable Limit of Activity for Collagen - Immobilized
           Lysozyme

(Venkatasubramanian, Bernath and Vieth 1973). The complex has
since retained this activity after four months of storage and intermittent use. Similarly, a glucose oxidase membrane made in the
same manner demonstrated an activity of 266 I.U./gram complex as
compared to 110 I.U./gram for a membrane prepared by electrocodeposition (Fernandes et al. 1973). The crosslinking agent apparently
"ties down" that fraction of the initial loading consisting of the
weakly bound enzyme that is normally washed off in its absence.
This is another example of collagen's versatile chemical nature.

Binding Mechanism

It should be emphasized at this point that although glutaraldehyde appears to increase the loading capacity of collagen, it is
not essential for the preparation of stable complexes. These results suggest that the primary mode of binding may be cooperative,
noncovalent, physico-chemical interactions including hydrogen
bonding, electrostatic linkages and hydrophobic interactions.
These are the same types of forces that produce very stable bonds
between subunits of multienzyme complexes and between proteins
and antibodies.

An analysis of collagen's microstructure (Figure 6) provides
some indication of the location of potential binding sites on the
carrier. The basic molecular unit of collagen is a triple stranded
helix (Ramachandran 1963), sometimes called tropocollagen, composed of three similar but not identical polypeptide chains.
This fundamental molecular unit is a rigid rod about 2,800 Å long,
14 Å in diameter and having a molecular weight of approximately
300,000 (Gross et al. 1954, Boedtker and Doty 1958, Rich and
Crick 1962). It is composed of 33% glycine, 25% proline and hy-
droxyproline and a distribution of 16 other amino acids. About
5% of each molecule is composed of a nonhelical polypeptide region located at the ends of the rods. These regions, called telopeptides, contain a higher concentration of polar and charged
amino acid residues than the remainder of the molecule (Rubin and
Stenzel 1969). Tropocollagen molecules aggregate spontaneously
under physiological pH, temperature and ionic strength into bun-
dles called microfibrils which are several hundred angstroms in
diameter and many microns in length. The microfibrils are formed
in such a fashion that there is a 9% overlap head to tail in linear polymers and an approximate quarter-stagger "offset" of ad-
jacent macromolecules (Hodge and Petruska 1963). This leads to
a 700 Å axial repeat period and to the presence of holes which
are regularly spaced along the axial length of the microfibril.
The dimensions of these holes are approximately 400 Å long by
15 - 25 Å in diameter.

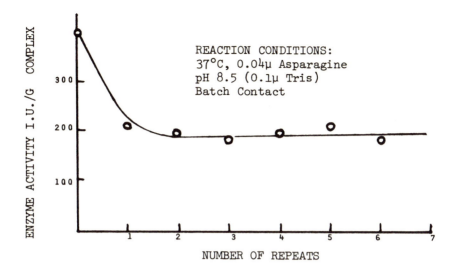

Figure 5. Stable Limit of Activity for Collagen - Immobilized
L-asparaginase

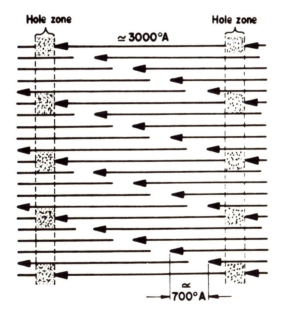

Figure 6. Collagen Microstructure

Since the hole regions of collagen microfibrils probably contain the randomly coiled telopeptides and since these telopeptides
contain a high density of polar and charged amino acid residues,
we postulate that these regions are the primary sites for enzyme
binding. We also postulate that binding takes place via cooperative, noncovalent, physico-chemical interactions. This mechanism
suggests the existence of a finite number of binding sites within
the carrier matrix. Figure 7 demonstrates that the amount of lactase impregnated by a collagen membrane (as evaluated by enzymatic
activity and tryptophan analysis) increases with increasing bath
enzyme concentration, approaching a saturation value asymptotically
at higher enzyme concentrations (Eskamani 1972). Similar behavior
has been observed for lysozyme, and Figure 8 shows that a linearized plot of the lysozyme data fits very closely to a Langmuirian
type sorption isotherm (Venkatasubramanian, Vieth and Wang 1972).

If the enzyme molecules are simply physically adsorbed or entrapped within the carrier matrix, then by prolonged washings with
buffer or salt solutions the enzyme may be detached from the carrier.
Figure 9 demonstrates the difference when subjected to salt washings between the behavior of collagen - catalase complexes formed
by simple adsorption and by complexation via cooperative, noncovalent interactions. In both cases a preformed collagen membrane

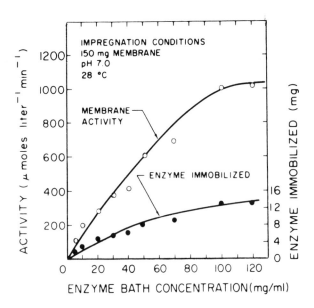

Figure 7. Sorption Isotherm for a Collagen - Lactase Membrane

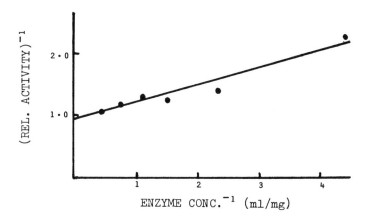

Figure 8.   Linearized Sorption Isotherm for a Collagen -
            Lysozyme Membrane

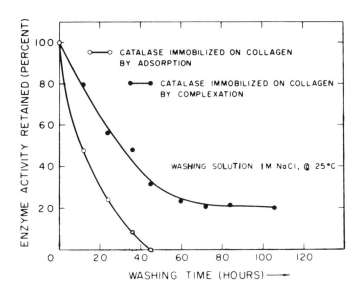

Figure 9.   Stability of Collagen - Catalase Membranes in
            1M NaCl

was immersed in the enzyme solution. In the case of adsorption
the membrane was not preswollen, nor was it dried before using.
In the case of complexation the membrane was preswollen, impregnated and then thoroughly dried before use. Initial activity of
the adsorbed membrane was 630 I.U./gram complex while that for
the complexed membrane was 7,200 I.U./gram. After 45 hours of
washing in 1M NaCl at 25°C. the membrane impregnated by adsorption
lost all of its activity. The complexed membrane retained 31%
of its activity after 45 hours and reached a stable limit of approximately 20% which remained constant after 60 - 100 hours of
washing. Furthermore, evidence for network formation in the complexed membrane was obtained by stress - strain measurements.
Table 5 shows that the molecular weight between network linkages
in the collagen matrix was decreased significantly by the complexation method and actually increased somewhat by simple adsorption.

If the binding of an enzyme to the collagen matrix occurs
through cooperative, noncovalent bonds, the interaction between
the two proteins should depend on the ionization states of the individual amino acid residues, which in turn depends on the impregnation pH. Figure 10 shows the variation in the amount of lysozyme immobilized (as measured by activity) as a function of impregnation bath pH (Venkatasubramanian, Vieth and Wang 1972).
The optimal pH for immobilization is between 8.0 and 8.5, which is
the median value of the isoelectric points of lysozyme (10.5 - 11.0)
and collagen (6.8). This implies that at this pH, where there is
a maximum amount of net charge difference on the two proteins, the
maximum complexation occurs. The pH-activity profile of free lysozyme is superimposed in Figure 10 to demonstrate that the observed effect is not due to a significantly greater activity of the
free enzyme at pH 8.0 - 8.5. For collagen - lactase complexes the
optimal pH for immobilization is about 7.2 which is also almost
midway between the isoelectric points of collagen and the enzyme
(Eskamani 1972).

Table 5

Network Properties of Collagen - Catalase Membranes

Film #	Time of Impregnation (hours)	Molecular weight between network linkages
Control	0	75,400
2	3	83,100
6	20*	54,300

* Complex was completely dried in a period of six hours.

In summary, experimental results demonstrate that collagen contains a finite number of enzyme binding sites. Results also provide supporting evidence for a binding mechanism of enzymes to collagen through multiple, noncovalent, physico-chemical inter- actions. It appears that the primary sites of binding are the telopeptides located within holes in the collagen microstructure. Glutaraldehyde treatment directly after immobilization may increase the membrane's stable limit of activity by covalently crosslink- ing the weakly bound enzyme molecules to lysine or arginine resi- dues in the helical region of the tropocollagen molecule.

Properties of Collagen - Enzyme Complexes

The properties of enzymes immobilized on collagen are generally not altered significantly although the effect varies somewhat for each enzyme. Figures 11 and 12, for example, show pH- and temper- ature-activity profiles, respectively, for free and immobilized lactase (Eskamani 1972). Immobilized lactase appears to have a slightly increased pH stability at both high and low pH's but has a decreased temperature stability at lower temperatures. The op- timum values in both cases are not affected, however. Figure 13

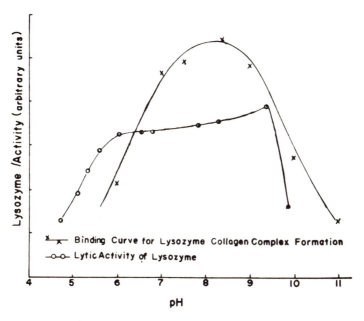

Figure 10. pH-Binding Curve for a Collagen - Lysozyme
            Membrane

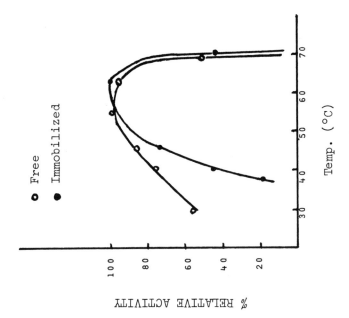

Figure 12.   Temperature-Activity Profile for
a Collagen – Lactase Membrane

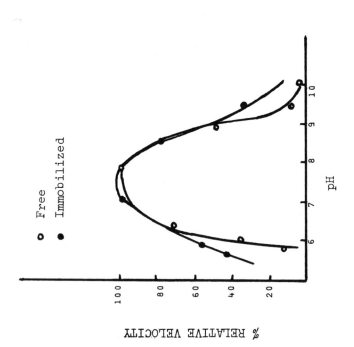

Figure 11.   pH-Activity Profile for a Collagen –
Lactase Membrane

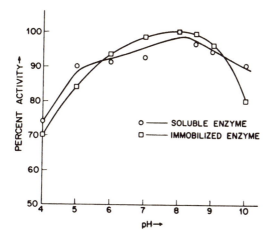

Figure 13.   pH-Activity Profile for a Collagen -
L-asparaginase Membrane

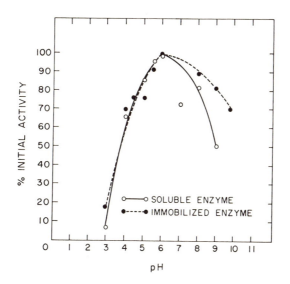

Figure 14.   pH-Activity Profile for a Collagen - Glucose
Oxidase Membrane

shows the similarities in the pH activity profiles of soluble and
immobilized l-asparaginase (Venkatasubramanian, Bernath and Vieth
1973). Figure 14 shows that immobilized glucose oxidase appears

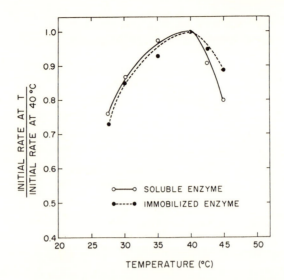

Figure 15.   Temperature-Activity Profile for Collagen -
Glucose Oxidase Membrane

to have increased pH stability at high pH's, and Figure 15 shows
the enzyme's temperature-activity profile (Constantanides et al.
1973). In both cases the optimum values are unaffected. The increased stability of the immobilized enzyme at higher bulk pH's
may reflect the formation of gluconic acid within the membrane
which would keep the microenvironmental pH somewhat lower than the
bulk.

The above results indicate that the microenvironmental pH of
the collagen matrix does not appear to be much different than the
bulk pH. Due to collagen's chemical nature, however, it should be
possible to design a desired microenvironmental effect. For example, by blocking amino or carboxyl groups or by adding additional
charged groups to the carrier the surface charge could be altered,
which in turn could affect the local pH. A highly negatively
charged membrane would attract hydrogen ions, maintain a lower
microenvironmental pH than the bulk solution, and thus stabilize
the enzyme in a highly alkaline substrate stream. A highly positively charged membrane would have the opposite effect. Of course,
the effect of such alterations on enzyme binding should always be
taken into consideration.

Michaelis constants for collagen-immobilized enzymes (as for
all immobilized enzymes) are highly dependent on external and internal diffusion effects. Figure 16 shows that the apparent K_m
for immobilized 1-asparaginase is 8.33 x10^{-4} M, some 80 times the
value for the soluble enzyme (Venkatasubramanian, Bernath and

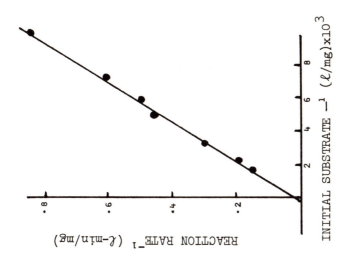

Figure 17. Lineweaver – Burk Plot for a
            Collagen – Lysozyme Membrane

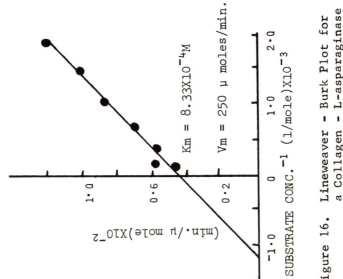

Figure 16. Lineweaver – Burk Plot for
            a Collagen – L-asparaginase
            Membrane

Vieth 1973). Figure 17 shows that the apparent K_m for immobilized lysozyme acting on an insoluble cell wall suspension is 6500 mg/l, compared to approximately 120 mg/l for the soluble enzyme (Venkatasubramanian, Vieth and Wang 1972). Immobilized glucose oxidase has an apparent K_m of 0.072 M which is only about 3.5 times the value for the soluble enzyme (Constantanides et al. 1973). The apparent K_m for immobilized lactase is 0.1 M compared to a value of 0.077 M for the free enzyme. These parameters actually have little value other than providing some gross indication of the existence of microenvironmental effects. If the design of large scale industrial enzyme reactors is to be successful, mass transfer and other microenvironmental effects must be considered on a rational basis as a total combination of elementary steps. This will be discussed in more detail later in this paper.

Biocatalytic Modular Reactor

In designing practical enzyme processes the development of a good carrier material solves only part of the problem. It is equally important to design an efficient reactor system to take advantage of the carrier's excellent properties. With this in mind we developed the spiral wound membrane reactor shown in Figure 18. This reactor, called a biocatalytic module, consists of a collagen - enzyme membrane layered onto a porous backing material, rolled into a spiral and placed into a tubular shell. The substrate stream can be pumped in axially through the capillaric channels that are formed by the spiral configuration or through the membrane under forced permeation conditions. The advantages

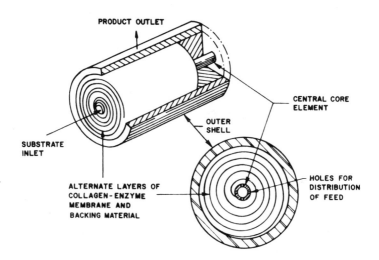

Figure 18. Biocatalytic Module

of this configuration include good mixing in the capillaric channels which decreases external diffusion effects, a thin membrane
which decreases internal diffusion effects, a low pressure drop
through the reactor, a large amount of catalytic surface area
within the reactor, a simple procedure for packing and unpacking,
and a resistance to channeling and compaction problems.

Most of our kinetic studies utilized the above reactor configuration in a recycle mode. For design purposes, however, we
have also studied biocatalytic modules operating as continuous
flow through reactors. Figure 19 shows the stability of a collagen - 1-asparaginase complex module operating in this mode. Under
the indicated conditions the module continuously converted 20% of
a 4mM substrate stream for a period of five days.

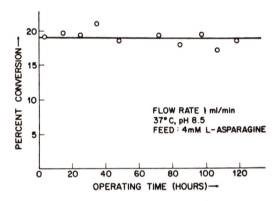

Figure 19.   Continuous Operation of a Collagen -
             L-asparaginase Module

Batch data from recycle reactors have been extremely helpful
in designing continuous processes. For example, Figure 20 shows
the performance of a recycle reactor containing collagen - immobilized whole cells which have glucose isomerase activity. The
cells, which are Streptomyces venzuelae, have been heat-treated to
inactivate undesirable proteins and to fix glucose isomerase to
the interior of the cell. The complex was made by macromolecular
complexation and glutaraldehyde treatment. A recycle reactor
which contains 27 units of glucose isomerase activity/ml. reactor
converts 50% of a 1 M dextrose stream in about 15 hours at 70°C.
and pH 7. We observe 40% conversion in about 3 hours. In designing a continuous reactor for 40% conversion we used a complex
which contributed 29 units/ml. reactor and we adjusted the residence time to 2.5 hours. The desired level of conversion was
achieved and maintained for a period of 15 days continuous operation as shown in Figure 21 (Vieth, Wang and Saini 1973). After
15 days part of the backing material disintegrated and the membrane overlapped in some areas, decreasing the effective catalytic

surface area of the reactor. This occurred again at 25 days, causing a further drop in conversion. After 40 days of operation the membrane itself was still intact. To prevent recurrence of this behavior we now use a more durable backing material, namely Vexar®, a reverse osmosis membrane support made by DuPont.

Mass Transfer Considerations

As mentioned earlier, apparent Michaelis constants are of little value in scaling-up and controlling immobilized enzyme reactors for commercial processes. These parameters are extremely dependent on mass transfer effects and thus on reactor size, configuration, throughput and conversion. It is therefore important to consider external diffusion and combined internal diffusion and reaction as a combination of elementary steps. Such an analysis would provide estimates of the mass transport coefficient through the fluid boundary layer, the effectiveness factor for the membrane and the uncoupled kinetic parameters of the system. These estimates can then be used to design a reactor for any combination of throughput and conversion. With this in mind we have developed the corresponding design equations for steady state (fixed-bed) and batch reactors containing membranes, capsules or beads. Furthermore, we have considered both reversible and irreversible reaction. As a first approximation we have considered first order, isothermal reactions which lead to analytical solutions for all of the cases mentioned above (Vieth et al. 1973).

The easiest case to apply, and one which serves as a first approximation for our system, is that for a steady state first order irreversible reaction in a membrane. At steady state for this system the substrate flux through the stagnant fluid layer surrounding the membrane must equal that into the membrane, i.e.,

(1) $J = K_L (C_b - C_s) = \eta L k C_s$

where J is the substrate flux in moles/sec.cm.^{2}, K_L is the external mass transfer coefficient in cm./sec., C_b and C_s are the bulk and surface concentrations of substrate, respectively, in moles/cm.^{3}, η is the effectiveness factor, k is the first order rate constant in sec.$^{-1}$ and L is the half-thickness of the membrane in cm. From this equation it follows that:

(2) $C_b/J = 1/K_L + 1/\eta k L = 1/K_o$

where K_o is an overall coefficient to account for diffusion and reaction. Using this overall coefficient in a steady state material balance on the substrate passing through a differential element of a fixed bed reactor gives:

(3) $dY_b/d\theta = - K_o a Y_b$

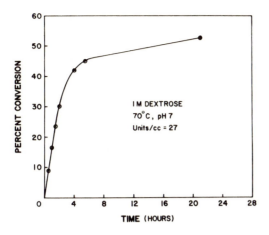

Figure 20.   Collagen - Immobilized S. venzuelae (glucose
            isomerase) in Batch Reactor

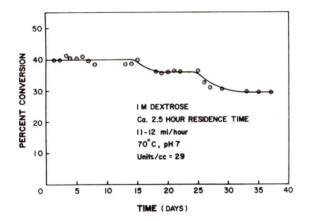

Figure 21.   Continuous Operation of Collagen - S. venzuelae
            Module

where Y_b is the fraction of unconverted substrate, θ is the residence time in sec., and a is the surface area per unit of reactor
volume in cm.$^{-1}$

The unconverted fraction of substrate can be determined experimentally for different values of the residence time. Knowing
these parameters it is then possible to calculate the combined
constant $K_o a$. Since the external mass transfer coefficient, K_L,

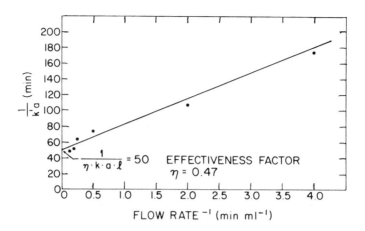

Figure 22.   Correlation of Mass Transfer Coefficient for a
             Collagen - Lactase Membrane

is a function of linear velocity, the combined coefficient should
be dependent on flow rate through the reactor. Furthermore, a
plot of $(K_O a)^{-1}$ versus reciprocal flow rate should yield a straight
line with an intercept equal to $1/\eta$ kaL, a factor independent of
flow rate. Figure 22 shows the results of the above analysis
applied to a collagen - lactase module (Eskamani 1972). Estimates
of the first order rate constant for the free enzyme, the catalytic
surface area per unit volume of reactor and the half-thickness of
the membrane allow the effectiveness factor, η , to be estimated.
The value of 0.47 suggests that about half of the bound enzyme is
not being totally utilized due to internal diffusion effects.
Figure 23 shows the same plot for lysozyme acting on its insoluble
cell wall substrate (Venkatasubramanian, Vieth and Wang 1972).
In this case the effectiveness factor is 0.04, inferring that only
4% of the bound enzyme is being used. This low value would be expected since the substrate's large size should prevent diffusion
into the membrane and limit reaction to the surface.

We are currently in the process of applying the combined design equations to data derived from more complex systems. One
such example is the isomerization of glucose to fructose by immobilized glucose isomerase in a reversible first order reaction.
Analyses of this type should greatly facilitate the design and
control of large scale commercial reactors. Future studies in our
laboratory will involve the design and operation of scaled-up reactors for a number of candidate enzyme systems with potential for
immediate industrial application.

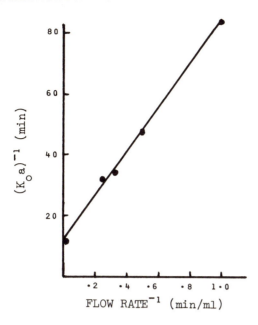

Figure 23. Correlation of Mass Transfer Coefficient for a
Collagen - Lysozyme Membrane

Related Studies

In an attempt to further improve contact between substrate
and enzyme in the modules we investigated the effect of pumping
substrate through the membrane under pressure. Figure 24 illustrates the results for the hydrolysis of sucrose by immobilized
invertase. The data indicate that the sucrose reaction is sub-
strate-inhibited. Immobilization of the enzyme moves the optimum
substrate concentration to slightly higher levels, probably due to
the difference between microenvironmental and bulk concentrations,
but the optimum reaction velocity is not altered significantly.
Pressure has a slight effect on the free enzyme. It would be assumed that the effect of pressure on the immobilized enzyme would
be to improve enzyme - substrate contact and thus follow more
closely the behavior of the free enzyme under pressure. As seen
in Figure 24, however, the optimum substrate concentration is
moved to still higher levels, i.e., substrate inhibition is de-
layed, and the optimum reaction velocity is increased by 40% over
the free enzyme (Venkatasubramanian and Vieth 1973). It is obvious that collagen, under the agency of hydrostatic pressure, has
in some way affected the reactivity and reaction path of the bound
enzyme. We are currently studying this phenomenon in detail.

Pumping a substrate stream through a membrane has other

advantages in addition to improving contact between substrate and enzyme. One of the most obvious is the possibility of achieving simultaneous reaction and rejection of undesirable components. A co-spiral wound module consisting of collagen - immobilized urease for urea conversion and a cellulose acetate membrane for salt rejection is currently under study in our laboratory. The objective is to produce a biocatalytic module which will aid in the production of potable water from urinous waste stream (Davidson et al. 1973). For some applications a collagen membrane alone may be sufficient to achieve both reaction and rejection. This characteristic of collagen membranes adds an additional level of flexibility to our reactor systems.

Interdisciplinary Program

In our attempts to design practical enzyme processes for areas where immediate socioeconomic need exists we realized that the expertise required is not generally found within a single academic department. For this reason we have established the Rutgers Interdisciplinary Enzyme Technology Group which consists of representatives from five academic departments within the University. Our research team consists of chemical/biochemical engineers, biochemists, microbiologists, food scientists, and bacteriologists. Individuals within the group contribute a high level of expertise in materials science, reactor design, optimization and control of fermentations and enzyme reactions; induction, repression and mutation in microbial production of enzymes; isolation, purification and characterization of enzymes; mechanistic studies, cell morphology, lytic enzymes, flavor evaluation and nutrition in foods and computer analysis of data. In addition, these individuals bring to the group contacts from a variety of industries and other organizations which enable us to evaluate the potential needs of industry and society and to facilitate the transfer of new processes after they are developed. This interdisciplinary approach has provided a broad awareness of the overall problem, a specific expertise in each of the various steps involved in the development of immobilized enzymatic processes and a group that can communicate with any potential user of our developments regardless of their field of specialization. Our experience and success over the last two years has demonstrated the tremendous advantages of an interdisciplinary program.

CONCLUSION

In conclusion, we have developed an enzyme carrier that is inexpensive and simple to fabricate, maintains enzyme stability, resists mechanical and microbial degradation, is highly penetrable, can be utilized with virtually all types of enzymes and whole cells

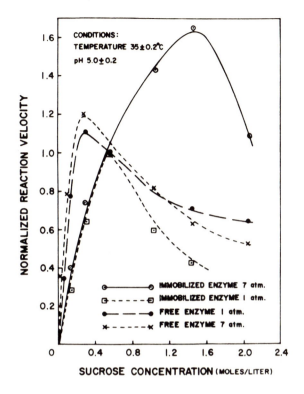

Figure 24.   Effect of Pressure on the Rate of Sucrose
             Hydrolysis by Collagen - Immobilized Invertase

and can be integrated into a variety of reactor configurations.
In addition, we have developed the corresponding design equations
for batch and continuous operations that account for mass transfer
effects associated with both bulk and internal diffusion. We have
done this because we believe that the indiscriminate utilization
of apparent bulk parameters in design or scale-up procedures will
not be very productive. If design of large scale industrial enzyme reactors is to be successful, mass transfer effects and other
microenvironmental effects must be considered on a rational basis
as a total combination of elementary steps. This line of reason-
ing has led to the development of a theory that can be utilized to
dependably predict and design the behavior of our system. These
points indicate that by following rational engineering principles
we have developed a novel process that is available for immediate
industrial applications.

ACKNOWLEDGMENT

The authors acknowledge with gratitude the support of the National Science Foundation, Grants-in-Aid GK-14075, GK-28866, and GI-34287. They also extend their thanks to Professors S. S. Wang, A. Constantanides and S. G. Gilbert; to Dr. A. Eskamani; and to Messrs. P. Fernandes, R. Saini and K. Venkatasubramanian for their contribution to this work.

REFERENCES

Boedtker, H. and P. Doty (1958). J. Amer. Chem. Soc. 80, 1269.

Bowes, J. H. and R. H. Kenten (1948). Biochem. J. 43, 369.

Constantanides, A., Vieth, W. R. and P. M. Fernandes (1973). Mol. and Cell. Biochem. 1, 127.

Davidson, B., Vieth, W. R., Wang, S. S. and S. Zwiebel (1973). Paper No. 17a, Sixty-Sixth Annual Meeting, AICh.E., Philadelphia, Pa., November 15.

Eskamani, A. (1972). Characterization of Lactase Immobilized on Collagen, Ph.D. Dissertation, Dept. of Food Science, Rutgers University, August.

Fernandes, P. M., Constantanides, A., Vieth, W. R. and D. Mody (1973). Chem. Engg. Sci. (in press).

Gross, J., Highberger, J. H. and F. O. Schmitt (1954). Proc. Nat. Acad. Sci., U.S., 40, 679.

Hochstadt, H. R., Park, F. and E. R. Lieberman (1960). U.S. Patent No. 2,920,000.

Hodge, A. J. and J. A. Petruska (1963). In Aspects of Protein Structure, G. N. Ramachandran, Ed., Academic Press, N.Y., 289.

Lehninger, A. L. (1970). Biochemistry, Worth Publishers, Inc., New York, 115.

Ramachandran, G. N. (1963). In Aspects of Protein Structure, G. N. Ramachandran, Ed., Academic Press, N.Y., 39.

Rich, H. and F. H. C. Crick (1962). J. Mol. Biol. 3, 483.

Rubin, A. L. and K. H. Stengel (1969). In Biomaterials, L. Stark and G. Agarwal, Eds., Plenum Press, N.Y., 157.

Venkatasubramanian, K., Bernath, F. and W. R. Vieth (1973). "The Use of Collagen-Immobilized Enzymes in Blood Treatment," Enzyme Engineering Conference, Henniker, N.H., August 10, (in press).

Venkatasubramanian, K. and W. R. Vieth (1973). Biotech. Bioeng. 15, 583.

Venkatasubramanian, K., Vieth, W. R. and S. S. Wang (1972). J. Ferm. Tech. 50, 600.

Vieth, W. R., Mendiratta, A. K., Mogensen, A. O., Saini, R. and K. Venkatasubramanian (1973). Chem. Engg. Sci. 28, 1013.

Vieth, W. R., Wang, S. S. and R. Saini (1973). Biotech. Bioeng. 15, 565.

Woessner, J. F., Jr. (1968). In Treatise on Collagen 2B, B. S. Gould, Ed., Academic Press, N.Y., 253.

THE IMMOBILIZATION OF ENZYMES WITH IMIDOESTER-CONTAINING POLYMERS

Oskar R. Zaborsky

Corporate Research Laboratories
Esso Research and Engineering Company
Linden, New Jersey 07036

INTRODUCTION

The most prevalent method for immobilizing water-soluble enzymes is by their covalent attachment to water-insoluble functionalized supports, a wide variety of which has been utilized (1). We now wish to report some of our studies on the use of water-insoluble methyl imidoester-containing polymers for this purpose.

The imidoester functionality has several desirable characteristics (2-4). Imidoesters react selectively with only the terminal α-amino group and the ϵ-amino groups of lysyl residues of proteins to form amidines (eq. 1). The resulting amidinated proteins

$$R-\overset{+}{\underset{OR}{C=NH_2}} + E-NH_2 \longrightarrow R-\overset{+}{\underset{NH-E}{C=NH_2}} + ROH \qquad (1)$$

exhibit negligible changes in their conformation from that of the native proteins and retain the same net overall charge. Amidinated proteins also exhibit an enhanced stability toward trypsin hydrolysis (4).

An additional feature of amidines, found to be very useful in modification studies, is the reversibility of the reaction. Removal of the amidino group from a protein can be accomplished by treatment of the modified protein with ammonia-ammonium acetate solution or with hydrazine (eq. 2). With a water-insoluble enzyme-polymer conjugate, ammonolysis would permit the reversible attachment of an enzyme. Likewise, in principle, the amidinated

$$R-C=\overset{+}{N}H_2 \; + \; NH_3 \rightleftharpoons R-C=\overset{+}{N}H_2 \; + \; E-NH_2$$
$$\underset{NH-E}{} \qquad\qquad\qquad \underset{NH_2}{} \qquad\qquad (2)$$

derivative could be exchanged with the same or different enzyme(s) (eq. 3). This reversibility, which is known for low molecular weight amines (5), would permit either the replacement of inactivated, bound enzyme with active, soluble enzyme or the simultaneous attachment of different enzymes, E_1, E_2, etc.

$$\underset{NH-E}{|\!-C=\overset{+}{N}H_2} \quad\xrightarrow{\;E'-NH_2\;}\quad \underset{NH-E'}{|\!-C=NH_2} \;\; + \;\; E-NH_2$$
$$\qquad\qquad\qquad\qquad\qquad\qquad\qquad\qquad (3)$$

(where $E'-NH_2$ is the active enzyme $E-NH_2$ or a different enzyme E_1, E_2, etc.).

Low-molecular-weight imidoesters exhibit varying degrees of stability in aqueous solutions (6-9); depending on the pH, hydrolysis produces esters, amides, or nitriles (eq. 4). However, even this apparent disadvantage can be viewed as being beneficial with

$$\underset{OR'}{R-C=\overset{+}{N}H_2} \quad\begin{array}{c} \overset{H^+}{\underset{H_2O}{\nearrow}} \quad \underset{OR'}{R-C=O} \; + \; NH_4^+ \\[2em] \overset{OH^-}{\underset{H_2O}{\searrow}} \quad \underset{NH_2}{R-C=O} \; + \; R'OH \end{array} \qquad (4)$$

or

$$R-C\equiv N \; + \; R'OH$$

regard to immobilization. The hydrolysis of any unreacted imidoester group of an enzyme-polymer conjugate would be desirable because a positively charged group is transformed into an uncharged group (imidoester to ester, amide or nitrile). Thus, non-specific adsorption of proteins or other charged compounds would be reduced. Charge elimination would occur only at the non-enzymatically modified positions of the polymer.

Imidoesters are also easily synthesized from a variety of inexpensive starting reagents (6). Most often, they are prepared from nitriles via the Pinner synthesis or from amides via suitable O-alkylations. The incentives for using nitrile-containing polymers

as the starting materials for the preparation of imidoester-containing polymers are likewise numerous. Nitrile-containing polymers
are available (commercially or synthesized) and inexpensive; they
have good mechanical, chemical and microbial stability; and they
are available in a variety of different forms (powder, fiber, film,
etc.). Further, a wide selection of microenvironments is available
for the to-be-immobilized enzyme because of the rather large
selection of nitrile-containing polymers (homopolymers, copolymers,
graft polymers or cyanoethylated supports). The preparation of
imidoester-containing water-insoluble polymers has been reported
previously by only one group of workers (10-12).

MATERIALS AND METHODS

Reagents employed for the immobilization of enzymes and the
characterization of the enzyme-polymer conjugates were of commercial
grade and were used without further purification unless so stated.
The enzymes trypsin (Code TRL) and α-chymotrypsin (Code CDI) were
obtained from Worthington Biochemical Corporation, Freehold,
New Jersey.

Preparation of Imidoester-Containing Polymers

The conventional Pinner synthesis employing hydrogen chloride
gas and methanol was used to prepare methyl imidoester-containing
polymers from the following parent nitrile-containing polymers of
powder or fiber form: polyacrylonitrile homopolymer, powder, gift
from A. F. Smith, of E. I. duPont de Nemours & Co., Textile Fibers
Department; acrylic carpet staple, Creslan(R), regular shrinkage,
merge No. 225808, type 61HB, gift from W. J. Bartlett of American
Cyanamid Co., Fibers Division; acrylic fiber, Chemstrand(R),
No. 7268086-2 A-16, gift from P. H. Hobson of Chemstrand Research
Center, Inc., Monsanto Company; styrene-acrylonitrile copolymer
(ca. 28% acrylonitrile) SAN RMD-4511 powder, gift from G. Webster
of Union Carbide Corporation, Chemicals and Plastics Division. The
procedure used was similar to that previously given by Tazuke
et al. (10-12). A typical preparation follows.

To a cooled (-10 to 5°C) and magnetically stirred suspension
of 10 g (0.188 mole nitrile) polyacrylonitrile (PAN) powder and
100 ml methanol (previously dried over 4A molecular sieves and
distilled) in a 250-ml 3-necked round-bottomed flask was slowly
introduced dry hydrogen chloride gas. The flask was equipped with
a low temperature thermometer and Drierite drying tube. Addition
of the HCl gas was carefully monitored because heat is evolved at
the initial stages of the reaction. After addition of the hydrogen
chloride gas (usually at least a ten molar excess, 68.5 g), the
mixture was allowed to come to room temperature and stirred for

2 to 3 d. The modified polyacrylonitrile powder (PANIE) was
filtered with a sintered glass funnel, washed thoroughly with
dried methanol and then anhydrous ether, and vacuum dried over
KOH flakes. The imidoester content of the resulting material was
determined by titration of the hydrochloride salt with base using
the automatic titrator, by potentiometric titration of the soluble
chloride with silver nitrate, or by infrared measurements.

Coupling of Enzymes to Imidoester-Containing Polymers

The coupling of various enzymes to the above-mentioned methyl
imidoester-containing polymers was conducted in the following
manner as described here for trypsin and PANIE powder. The automatic
titration assembly described later in detail (Assay of Soluble and
Immobilized Enzyme Section) was used to maintain a constant and
preset pH.

A freshly prepared solution of trypsin (100 mg dissolved in
10 ml 0.1 $\underline{M}$ CaCl$_2$) was placed into a titration vessel having an
open side arm and equipped with a magnetic stirring bar. The pH
of the solution was adjusted to 9.0, and 1.0 g of PANIE powder was
added portionwise at room temperature keeping the pH constant by
continued addition of base. After the final addition of the activated polymer (requiring about 20 to 25 min), the mixture was
stirred for 2 to 3 h at room temperature. The suspension was then
filtered with a Millipore filtration unit (0.45 μm filter) and the
enzyme-conjugate was thoroughly washed with 0.1 $\underline{M}$ CaCl$_2$, water,
and various buffer and salt solutions until no enzymic activity
could be detected further in the wash. Washing of the enzyme
conjugates free of soluble enzyme was troubleless; the support is
non-porous in nature (preventing severe occlusion) and it does not
clog the filter.

The resulting enzyme conjugates were stored in buffers or in
water (for trypsin, usually in a 50 m$\underline{M}$ Tris-HCl buffer, pH 7, in
50 m$\underline{M}$ CaCl$_2$).

Protein Determinations

The protein content of the enzyme conjugates was determined by
amino acid analysis (after appropriate hydrolysis in 6 $\underline{N}$ HCl at
110°C for 20 h, evacuated sample), using a modification of the gas-
liquid chromatographic procedure described by Gehrke (13-15) and
co-workers. No interference was encountered in the preparation or
gas-liquid chromatographic analysis of the $\underline{N}$-trifluoroacetyl $\underline{n}$-butyl
ester derivatives of amino acids of interest (alanine, valine or
glutamic acid). Usually, the protein content of the immobilized
enzyme was determined from the amount of alanine experimentally
found and from the number of residues per molecule. Soluble

protein was determined spectrophotometrically at 280 nm using the
reported extinction coefficients for trypsin and α-chymotrypsin (16).

Assay of Soluble and Immobilized Enzymes

The assay of soluble and immobilized trypsin and α-chymotrypsin
was performed either spectrophotometrically or titrimetrically
using the substrates p-toluenesulfonyl-L-arginine methyl ester
(TAME) or benzoyl-L-tyrosine ethyl ester (BTEE), respectively (16).
Spectrophotometrically, soluble and immobilized enzyme derivatives
were assayed using a Beckman Kintrac VII A or Acta V spectrophotometer.
These instruments are equipped with a built-in magnetic stirring
system (Teflon-coated stirring bar inside the cuvette) and are
suitable for monitoring the rate of reaction of water-insoluble
enzyme derivatives, provided, of course, that the spectral interference due to the support is tolerable. The typical assay entailed
adding 5-50 μl of enzyme solution or suspension to a previously
equilibrated solution (at 25°C) of the standard assay mixture and
monitoring the absorbance change with time. A stirring speed well
above the minimum required for obtaining constant specific activities
of the immobilized enzyme conjugates was used throughout these
studies. Titration assays were conducted with a London Co. automatic
titration assembly at 25°C, which consisted of the following
components: a pH M-26 pH meter, a TT11 titrator controller, a
SBR2 titrigraph recorder, a TTA31 micro-titration assembly or a
TTA3 macrotitration assembly with thermostated cell compartments
and either combination electrodes of G222C and K4112 or G202C and
K401, respectively. A nitrogen stream was used to prevent carbon
dioxide absorption.

Thermal Inactivation of Soluble and Immobilized Enzymes

The following procedure was employed for determining the
thermal inactivation of soluble and immobilized trypsin and
α-chymotrypsin. A 1 to 2 ml sample of either soluble (containing
ca. 0.05-0.20 mg protein/ml solution) or immobilized
(containing ca. 10 mg immobilized enzyme conjugate/ml suspension)
enzyme in buffer was placed in a small test tube having a sealable
standard ground-glass joint. The closed test tube containing the
suspension or solution was then placed into a constant temperature
water bath of the desired temperature, and samples were removed
periodically and tested for enzymic activity. Before removing a
sample of immobilized enzyme, the suspension was vigorously shaken
with a Vortex mixer to ensure homogeneity of the sample. Duplicate
assays were performed at each time interval. The relative activities
reported were determined using the time at zero (no heating) as
100% activity for both the soluble and immobilized enzymes. All
assays were conducted at 25°C.

Lyophilization Stability

A 1-ml suspension of PANIE-trypsin (ca. 30 mg/ml in 0.05 M Tris-HCl buffer, pH 6.89) was placed in a vial and lyophilized at -10°C with a Virtis automatic freeze dryer (controlled shelf temperature) model 10-010-TVG-24M. The lyophilized material was resuspended in water and its activity and absorbance change/min/mg of immobilized enzyme conjugate (ΔA/min/mgIE) determined.

Ammonolysis

To 5 ml ammonia-ammonium acetate solution (prepared from concentrated ammonium hydroxide and glacial acetic acid, 15:1, v/v), pH 11.1 in a vial was added ca. 100 mg of PANIE-trypsin, and the suspension was stirred gently at 5°C for 3 d. The mixture was filtered with a Millipore filtration unit, 0.45 μm filter, and the solid was washed with additional ammonia solution and then with water until no enzymic activity could be detected further in the wash. The activity of the fractions (soluble and immobilized) was determined. The specific activity and protein content of the recovered PANIE-trypsin were compared to those of the starting immobilized trypsin.

RESULTS AND DISCUSSION

Preparation of Imidoester-Containing Polymers

Various methyl imidoester-containing polymers were prepared from the corresponding nitrile polymers using methanol and HCl gas via the Pinner synthesis. As expected and also previously reported (10-12), the extent of modification is increased at higher temperatures and longer reaction times. Under the conditions described in the Materials and Methods Section using methanol as the alcohol, the extent of modification for the polyacrylonitrile powder was usually 2 to 4%. The extent of modification of the polymers determined by either titration of the hydrochloride salt or by titration of the soluble chloride agreed within experimental error. The infrared spectra of the starting PAN powder and the resulting PANIE are given in Figs. 1 and 2. The strong absorption bond at 1680 cm^{-1} is assigned to the C=N stretching frequency of the imidoester; the weak band at 1230 cm^{-1} is assigned to the C-O stretching frequency of the C-O-CH$_3$ group of the ester. Although the modification is rather low (2 to 4%), it should be kept in mind that this value represents only surface modification of the polymer. Polyacrylonitrile does not swell in methanol. Likewise, the lower modification obtained with fibers (usually < 1%) is the result of an even lower surface area.

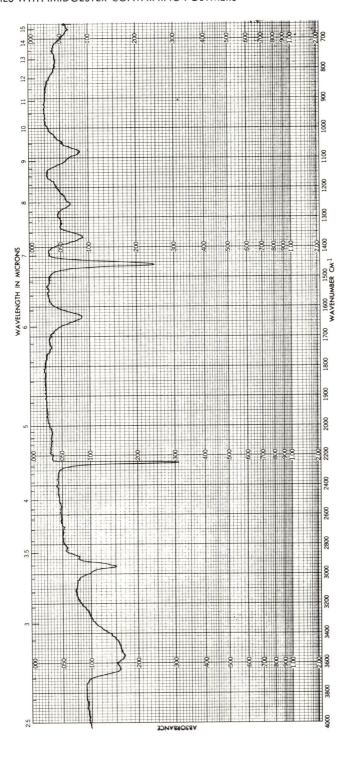

Figure 1. Infrared spectrum of polyacrylonitrile powder (PAN) obtained on a Beckman IR 12 spectrophotometer with the KBr-disk technique.

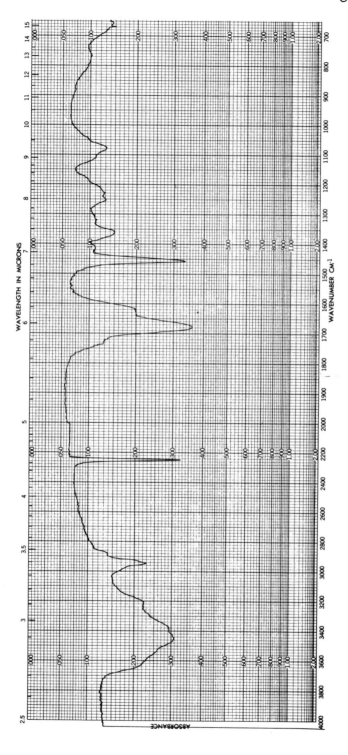

Figure 2. Infrared spectrum of methyl imidoester of
polyacrylonitrile powder (PANIE) obtained
on a Beckman IR 12 spectrophotometer with
the KBr-disk technique.

Cosolvents can be used to swell these nitrile-containing polymers (DMF for PAN and CH_2Cl_2 for SAN) and when used, the extent of modification is increased. However, the reprecipitated polymers are cumbersome to manipulate, and no further work was done along this line. The use of diethyl ether as a cosolvent (the usual solvent employed for the preparation of low-molecular-weight imidoesters) offers no distinct advantage with PAN.

Stability of PANIE

Because of the facile hydrolysis and thermal decomposition of low-molecular-weight imidoesters (6-9), the polymeric PANIE was stored initially at 5°C in a vacuum desiccator over KOH flakes or P_2O_5 pellets. However, based on the nearly constant amount of protein binding with a given PANIE preparation stored for a given period of time, it became evident that little, if any, decomposition occurred. Infrared spectral analysis confirmed this beyond a doubt. A comparison of the spectra of a PANIE powder stored at 5°C in a desiccator and taken two years apart revealed no changes in the intensity of the peaks or their positions. No decomposition to the methyl ester (appearance of 1735 cm^{-1} band) occurred during this time. Storage of the PANIE powder at room temperature in a desiccator for 3 to 4 months likewise revealed no changes in the spectrum. Even when stored at room temperature in the open for 4 d, no changes were evident. Imidoester-containing polymers, in contrast to low-molecular-weight imidoesters, appear to be rather stable over prolonged periods of time.

The hydrolytic stability of PANIE powder was additionally tested. The spectrum of a PANIE powder suspended in water for 10 min (then filtered and washed with methanol and ether and vacuum dried over P_2O_5) was identical to the spectrum of the untreated material. Further, the spectrum of a PANIE powder suspended in water for 4 d at room temperature (pH of the suspension being 3.78), likewise revealed no changes (no appearance of a stronger ester band of 1735 cm^{-1}). Treatment of the PANIE powder for 4 d at room temperature with 0.5 $\underline{M}$ $NaHCO_3$ (pH of suspension, 8.30) or with 0.5 $\underline{M}$ Na_2HPO_4 (pH of suspension, 9.09) followed by spectral analysis revealed no changes in the spectrum of either sample that could be attributed to decomposition of the imidoester functionality to ester or to the amide.

Coupling of Enzymes to Imidoester-Containing Polymers

The coupling of the enzymes trypsin and α-chymotrypsin to the imidoesters-containing polymers was performed at alkaline pH (from 8.5 to 10) for 0.5 to 5 h at room temperature. As expected, the amount of loading increased with increasing pH of the coupling solution and with the time of coupling (with a given batch of imidoester polymer). Typical protein binding for these enzymes ranged from 1.5 to 8 mg of enzyme/g of enzyme-polymer conjugate. Specifically, the protein binding for trypsin on the PANIE powder coupled at pH 9.5 for 0.5 h and coupled at pH 10.0 for 5 h was 2.3 and 6.8 mg trypsin/g of enzyme-PANIE conjugate, respectively. Again, it should be kept in mind that the PANIE polymers have been modified only at the surface and consequently low protein binding is anticipated. Optimization of the binding conditions was not investigated. Other enzymes coupled to imidoester-containing polymers, besides trypsin and α-chymotrypsin, included ribonuclease A and phenolase.

Activity of Bound Enzymes

The activities of the immobilized enzymes were measured using the standard assays with the Beckman spectrophomometers or the titration assembly. In both systems, a sufficient stirring speed was employed to ensure good mixing of the enzyme-polymer conjugates. With the spectrophotometric systems, no apparent deviation from linearity of absorbance change per time (initial), $\Delta A/min$, was observed under normal conditions, and no problems with reproducibility were encountered. Good activities were observed for all of the enzyme polymer conjugates examined in detail. The specific activity of an α-chymotrypsin-PANIE powder conjugate coupled at pH 9.5 for 0.5 h and having 1.5 mg protein/g of enzyme-polymer conjugate was 57 units/mg of bound enzyme. The specific activity of the corresponding native soluble enzyme was 49 units/mg of enzyme. This slight, but apparently real, activation of the immobilized enzyme was also exhibited by trypsin conjugates. The specific activity of the trypsin-PANIE powder conjugates prepared from various PANIE preparations and under different conditions gave surprisingly, but consistently, higher values than the native, soluble enzyme. For example, the specific activity of the trypsin conjugate coupled at pH 10.0 for 5 h having a protein content of 6.8 mg/g of enzyme-polymer conjugate exhibited a specific activity of 250 units/mg enzyme. The specific activity of the native, soluble enzyme was 210 units/mg of trypsin. Although many experimental errors could be invoked as the cause of this apparent anomaly, a careful check of the system (from hydrolysis to amino acid analysis) revealed no apparent technical difficulty. Although not specifically implicated, the most probable source of error in the specific activity determination is the amino acid analysis, i.e., the determination of the protein bound to the polymer. A loss in the amount of the amino acids used for protein determination caused possibly by decomposition during hydrolysis, non-uniform adsorption onto glass surfaces, or incomplete derivatization of the amino acids into the volatile derivatives could account for this observation. This would be an especially severe problem when analyzing samples of low protein content. However, in defense of the gas-liquid chromatographic analysis, it should be stated that this procedure provides for very sensitive, nanomole detection of amino acids. Such a sensitive method of analysis was especially suitable and needed for our present imidoester polymers which bind enzymes only to a low degree at their surface. The reason for the enhanced activity of the PANIE-conjugates of trypsin and α-chymotrypsin toward the low-molecular-weight substrates is still obscure.

Thermal Stability

The thermal inactivation of soluble and immobilized α-chymotrypsin and trypsin was determined at several temperatures. The results obtained for α-chymotrypsin at 50°C and for trypsin at 50 and 70°C are given in Figs. 3 to 5. At these temperatures, thermal inactivation is presumably favored over autolysis (17). As can be seen, the thermal stability of the water-insoluble enzyme conjugates at these temperatures is greater than that of the native, water-soluble enzymes. A biphasic behavior seems to be especially evident

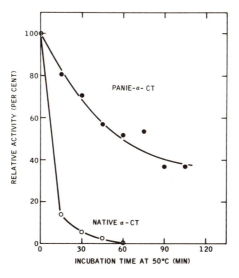

Figure 3.   Thermal inactivation of soluble and immobilized α-chymotrypsin at 50°C. Relative activities of soluble (0——0) and immobilized (●——●) enzymes were determined at 25°C employing the standard assay. Specific activity at time zero (no heating) was taken as 100% relative activity. Native α-chymotrypsin concentration was 200 μg/ml in 50 mM Tris-HCl buffer, pH 7.03. Same buffer for PANIE-$\overline{\alpha}$-chymotrypsin.

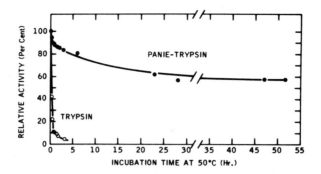

Figure 4.   Thermal inactivation of soluble and immobilized trypsin
             at 50°C.   Relative activities of soluble (O——O) and
             immobilized (●——●) enzymes were determined at 25°C
             employing the standard assay.   Specific activity at
             time zero (no heating) was taken as 100% relative
             activity.   Native trypsin concentration was 50 μg/ml
             in 0.1 M̲ Tris-HCl buffer, pH 7.08 in 0.1 M̲ NaCl.   Same
             buffer for PANIE-trypsin.

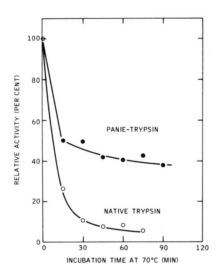

Figure 5.   Thermal inactivation of soluble and immobilized trypsin
             at 70°C.   Relative activities of soluble (O——O) and
             immobilized (●——●) enzymes were determined at 25°C
             employing the standard assay.   Specific activity at time
             zero (no heating) was taken as 100% relative activity.
             Native trypsin concentration was 50 μg/ml (in 0.1 M̲
             Tris-HCl buffer, pH 7.08 in 0.1 M̲ NaCl).   Same buffer
             for PANIE-trypsin.

in Fig. 4 and 5 which show the inactivation of trypsin and PANIE-
trypsin. At 50°C, the PANIE-trypsin conjugate loses <u>ca</u>. 15% of its
initial activity in 1 h and in the same time span, the soluble
trypsin loses well over 90%. After this initial drop in activity
of the PANIE-trypsin, the conjugate loses activity only slowly, with
the loss approaching almost zero at greater than 25 h. The sudden
drop in activity is even more pronounced at 70°C. At this temperature the PANIE-trypsin conjugate loses <u>ca</u>. 50% of its initial
activity in 15 min; a greater loss (75%) occurs with the soluble
trypsin.

     An interpretation of this biphasic behavior which has exper-
imental support from both our work (18) and that of others (19,20)
is as follows. The reaction of the activated polymer with an
enzyme produces a mixture of modified enzyme molecules differing
in site and degree of substitution, conformational stability, etc.
<u>A priori</u>, as with any other chemical modification of a protein
which produces such a mixture, some of these enzyme derivatives
exhibit superior stability while others exhibit inferior or
unchanged conformational stability. At these temperatures of 50
and 70°C, rapid thermal degradation of the labile isomers occurs
with a sudden drop in activity and a thermally more stable population remains. In essence, degradation of the thermally labile
isomers of the mixture gives an enzyme-polymer conjugate which
exhibits superior thermal stability relative to the native, soluble
enzyme. Although the precise chemical nature for the increased
thermal stability is not known at this time, appropriately
positioned crosslinks between the enzyme and the polymer chains of
the support (both intra- and intermolecular crosslinks) could be
responsible for the observed behavior. Appropriately positioned
<u>intramolecular</u> crosslinks in <u>soluble</u> enzyme derivatives can affect
enhanced conformational stability (18). Gabel has similarly
expressed this possibility to account for the enhanced thermal
stability observed in Sepharose-enzyme conjugates (19).

                          Storage Stability

     Good storage stability was qualitatively noted for all the
PANIE enzyme conjugates. A quantitative study of PANIE-trypsin
revealed excellent storage stability for this conjugate. Within
experimental error, no loss of activity was observed even after a
year of storage at 5°C in 0.1 <u>M</u> phosphate buffer, pH 6.7.

                       Lyophilization Stability

     No detectable inactivation occurred during the lyophilization
of the PANIE-trypsin conjugate. The ΔA/min/mgIE values before and
after lyophilization were identical (0.165).

Solvent Stability - A Qualitative Observation

The preparation of samples for amino acid analysis involves washing the enzyme conjugates with buffer, water, methanol and ether and then drying the residue in a vacuum desiccator over P_2O_5. Such solvent treated PANIE-trypsin, when resuspended in water exhibited surprisingly substantial activity. However, no quantitative assessment was made.

Enzyme "Leakage"

The "leakage" of an enzyme from a support can be caused by physical desorption or by covalent bond breakage of the connecting bond between the enzyme and the support or between atoms of the support (degradation and solubilization of the support material). Periodic checks of the filtered storage solution (enzyme conjugates stored in various buffers or water) of the PANIE conjugates revealed no enzymic activity. A storage solution of a PANIE-trypsin derivative (stored at 5°C in 50 mM Tris-HCl buffer, pH 6.89) exhibited no detectable activity even after two years. Evidently, no degradation (solubilization) of the support seems to occur.

Ammonolysis

No enzymic activity was detected in the ammonia exchange solution (the original filtrate) or the subsequent wash solutions. The treated PANIE-trypsin conjugate did, however, show substantial activity. Analysis of the conjugate revealed the solid to have a $\Delta A/min/mgIE$ value of 0.149 ($\Delta A/min/mgIE$ of starting trypsin derivative was 0.164) and an mgE/mgIE value of 1.98×10^{-3} (mgE/mgIE value of starting trypsin derivative was 2.34×10^{-3}). See Table I. The specific activity of the recovered and starting

Table I

Reversibility Test of PANIE-Trypsin
With NH_4OH - NH_4OAc

Activity or Binding	Untreated	NH_4OH - NH_4OAc Treated	Percentage Change
$\Delta A/min/mgIE$	0.164	0.149	9.1 loss
mgE/gIE	2.34	1.98	15.4 loss
Specific Activity (units/mgE)	398	417	4.8 increase

PANIE-trypsin conjugates was 417 and 398 units/mgE respectively. The reduction of the ΔA/min/mgIE value from 0.164 to 0.149 (a 9.1% loss of activity caused presumably by the removal of protein) and the reduction of the mgE/mgIE value from 2.34 x 10^{-3} to 1.98 x 10^{-3} (a 15.4% loss of actual protein as measured by amino acid analysis) suggest that some trypsin was indeed removed from the conjugate. In contrast, the specific activities of the samples tested were essentially equivalent within experimental error (4.8% difference, with the treated sample being more active). Although these results suggest the anticipated result, they are by no means unequivocal. A sample of PANIE-trypsin having a higher protein content must be used to test for this reversibility.

A finding of a more favorable imidoester system (i.e., a higher protein-containing immobilized system) substantiates this preliminary observation (O. Zaborsky, unpublished result). An immobilized α-chymotrypsin conjugate, prepared by adsorption of the enzyme onto porous silica and then crosslinking the adsorbed enzyme molecules with the bifunctional imidoester, dimethyl adipimidate (DMA), was subjected to the same ammonolysis treatment. The results of this study, summarized in Table II, substantiate the trend observed in the PANIE-trypsin study. The treated sample has a lower ΔA/min/mgIE value (a 71% loss) and a lower mgE/mgIE value (97% loss). In contrast, the specific activity of the treated sample has a dramatically higher value than the untreated derivative. This increase can be attributed to the removal of extra protein layers of the enzyme conjugate creating a more effective enzyme derivative. Evidently, diffusional restrictions could cause the high protein-containing derivative to be less effective than the lower modified derivative. Similar results have been observed by others with other immobilized systems (21).

Table II

Reversibility Test of
Porasil-α-Chymotrypsin-DMA with NH$_4$OH - NH$_4$OAc

Activity or Binding	Untreated	NH$_4$OH - NH$_4$OAc Treated	Percentage Change
ΔA/min/mgIE	0.0202	0.00587	71.0 loss
mgE/gIE	55.3	1.56	97.1 loss
Specific Activity (units/mgE)	1.14	11.3	991 increase

In summary, imidoester-containing polymers are good supports for immobilizing enzymes and, in principle, have certain character- istics which make them potentially even superior to presently available ones.   They are easily prepared from nitrile-containing polymers (polyacrylonitrile being the simplest nitrile polymer) via the Pinner synthesis using methanol and hydrogen chloride gas, and they are more stable toward hydrolysis than low-molecular-weight esters.   Imidoesters react selectively with only the terminal α-amino group and the ϵ-amino groups of lysyl residues of proteins to form amidines.   Coupling of an enzyme to a polymer is conducted at slightly alkaline pH, and protein binding up to ca. 8 mg of enzyme/g of enzyme-polymer conjugate has been obtained.   Bonded trypsin and α-chymotrypsin exhibit excellent enzymic activity toward low-molecular-weight substrates and show an enhanced thermal stability.   The trypsin conjugate derived from the imidoester of polyacrylonitrile shows no loss of activity upon lyophilization or storage in buffer at 5°C for one year.

Future work in this area should include optimizing the synthesis of the imidoesters and the enzyme-polymer conjugates and extending the scope of the method, i.e. the use of more different polymers and proteins.

ACKNOWLEDGMENT

The author acknowledges the excellent technical assistance of Miss Jacqueline Ogletree.

REFERENCES

1.   Zaborsky, O. R. (1973) in Immobilized Enzymes, Chemical Rubber Co. Press, Cleveland.
2.   Hunter, M. J. and Ludwig, M. L. (1972) in Meth. in Enzymol. 25 Part B (Hirs, C. H. W. and Timasheff, S. N., ed.) 585-596, Academic Press, New York.
3.   Hunter, M. J. and Ludwig, M. L. (1967) in Meth. in Enzymol. 11 (Hirs, C. H. W., ed.) 595-604, Academic Press, New York.
4.   Hunter, M. J. and Ludwig, M. L. (1962) J. Am. Chem. Soc. 84, 3491-3504.
5.   Shriner, R. L. and Neumann, F. W. (1944) Chem. Rev. 35, 351-425.
6.   Roger, R. and Neilson, D. G. (1960) Chem. Rev. 61, 179-211.
7.   Okuyama, T., Pletcher, T. C., Sahn, D. J. and Schmir, G. L. (1973) J. Am. Chem. Soc. 95, 1253-65.
8.   Chaturvedi, R. K. and Schmir, G. L. (1968) J. Am. Chem. Soc. 90, 4413-4420.

9. Pletcher, T. C., Koehler, S. and Cordes, E. H. (1968) J. Am. Chem. Soc. 90, 7072-7076.
10. Tazuke, S., Hayashi, K. and Okamura, S. (1965) Kobunshi Kagaku 22, 259-263.
11. Okamura, S., Tazuke, S. and Hayashi, K. (1965) Japan 23,355 (to Research Foundation for Practical Life) C.A. (1966) 64, 3800 h.
12. Tazuke, S., Hayashi, K. and Okamura, S. (1965) Makromol. Chem. 89, 290-291.
13. Zumwalt, R. W., Roach, D. and Gehrke, C. W. (1970) J. Chromatogr. 53, 171-193.
14. Roach, D. and Gehrke, C. W. (1969) J. Chromatogr. 44, 269-278.
15. Gehrke, C. W. and Stalling, D. L. (1967) Sep. Sci. 2, 101-138.
16. Worthington Enzyme Manual (1972) Worthington Biochemical Corporation, New Jersey.
17. Kumar, S. and Hein, G. E. (1970) Biochemistry 9, 291-297.
18. Zaborsky, O. R. (1973) submitted for publication, Second International Enzyme Engineering Conference.
19. Gabel, D. (1973) Eur. J. Biochem. 33, 348-356.
20. Wang, J. H.-C. and Tu, J.-I. (1969) Biochemistry 8, 4403-4410.
21. Axen, R. and Ernback, S. (1971) Eur. J. Biochem. 18, 351-360.

BASIC CONCEPTS IN THE EFFECTS OF MASS TRANSFER ON IMMOBILIZED

ENZYME KINETICS

Bruce K. Hamilton, Colin R. Gardner*, and C.K. Colton**

Department of Chemical Engineering
Massachusetts Institute of Technology
Cambridge, Massachusetts 02139

ABSTRACT

The observed kinetics of immobilized enzymes can often be
influenced by mass transfer effects. To ensure straight-forward
and useful analysis of data, these influences should be anticipated
at the early stages of design and execution of kinetic experiments.
With adequate care, it is then possible to interpret meaningfully
values of commonly reported immobilized enzyme kinetic parameters,
such as the "apparent Michaelis constant," even if severe mass
transfer effects are present when rate data are obtained.

INTRODUCTION

Methods for analyzing soluble enzyme kinetics are well
established (Dixon and Webb, 1964; Segel, 1968; Plowman, 1972),
and comparative advantages of various approaches have been extensively assessed (Coleman, 1965; Dowd and Riggs, 1965; Cleland,
1967). Methods for analyzing immobilized enzyme kinetics, however,
are not so well systematized, even though the influence of mass-
transfer on biological kinetics has been recognized for many years
(e.g., Roughton, 1932, 1952, 1959). Recently, numerous theoretical
studies of the mediation of immobilized enzyme kinetics by mass
transfer effects have been presented, in some cases in conjunction
with experimental studies. The purpose of this paper is to briefly
review the basic mass transfer phenomena which can influence im-

* Department of Chemistry, University of Aberdeen, Aberdeen,Scotland.
**To whom correspondence should be addressed.

mobilized enzyme kinetics and to outline the various methods which have been proposed in the literature for analyzing such phenomena.

Basic concepts will be illustrated by analysis of a simple example, and then means for determining values of intrinsic kinetic parameters of immobilized enzymes will be highlighted. The need to evaluate intrinsic parameters arises from two sources: first, from the question as to whether intrinsic kinetic constants, such as the Michaelis constant, for an immobilized enzyme differ from those for the native form of the same enzyme free in solution, and second, from the desire to formulate accurate and versatile models for application in design of immobilized enzyme reactors. In particular, it will be shown how an appropriately determined "apparent Michaelis constant," which might be measured even with very significant mass-transfer effects present, can be used to calculate an intrinsic Michaelis constant for an immobilized enzyme. For perspective, some mention of electrostatic effects, which can also influence immobilized kinetics, will be made.

EXAMPLE MATHEMATICAL MODEL

The Catalytic Slab or Membrane

<u>Governing differential equation and boundary conditions</u>. For the purpose of illustration, consider a porous matrix in the form of a one-dimensional slab or membrane of thickness L in which enzymatic activity is uniformly distributed. Assume that the intrinsic kinetics of the immobilized enzyme follow the simple irreversible Michaelis-Menten expression:

$$v = \frac{V_m s}{K_m + s} \tag{1}$$

where v is the reaction rate per unit volume of porous catalyst, V_m is the maximum reaction rate per unit volume of porous catalyst, K_m is the intrinsic Michaelis constant for the immobilized enzyme, and s is the local substrate concentration within the voids of the porous material. Equation (1) is usually derived for soluble enzymes by invoking the "pseudo-steady state hypothesis" which holds when enzyme concentration is much less than substrate concentration (Heineken, <u>et al</u>., 1967). However, Aris (1972) has shown that, for immobilized enzymes, Equation (1) holds even if the enzyme concentration is very high, as have Engasser and Horvath (1973) for the case of steady state reaction.

We examine here the case where one surface of the membrane is impermeable to substrate and the other is maintained at a uniform concentration, s_s. It is assumed that substrate diffusion within the membrane can be represented by Fick's first law in terms of a concentration-independent effective diffusivity D_{eff}, that there are no interactions (e.g., electrostatic effects) between substrate and porous support, and that the values of V_m and K_m are the same for all immobilized enzyme and do not very throughout the matrix (as a result, for example, of varying pH or ionic strength). At steady state, a mass balance over a differential volume element of depth dx (Petersen, 1965; Satterfield, 1970) yields:

$$D_{eff} \frac{d^2 s}{dx^2} - \frac{V_m s}{K_m + s} = 0 \qquad (2)$$

The relevant boundary conditions are:

$$\text{I.} \quad s = s_s \quad \text{at } x = 0 \qquad (3)$$

$$\text{II.} \quad \frac{ds}{dx} = 0 \quad \text{at } x = L \qquad (4)$$

The solution to this problem in terms of the effectiveness factor (see below) is conveniently represented in terms of two dimensionless groups: a dimensionless Michaelis constant, $\nu = K_m/s_s$, and the modified Thiele modulus, ϕ_m, defined by:

$$\phi_m = L \left[\frac{V_m}{K_m D_{eff}} \right]^{1/2} \qquad (5)$$

Alternate moduli. Use of the modulus ϕ_m as defined above, in combination with the dimensionless Michaelis constant ν, permits separation of concentration- and non-concentration-dependent parameters. Various other moduli, however, can and have been employed, including a general asymptotic modulus (Bischoff, 1965; Moo-Young and Kobayashi, 1972; Horvath and Engasser, 1973), and other concentration dependent moduli (Fink, et al., 1973; Miyamoto, et al., 1973). The modulus ϕ_m will be used throughout this paper; others who have taken this approach include Thomas, et al. (1972) and Marsh, et al. (1973).

More complex kinetics. Although only the case of a single reaction promoted by an immobilized enzyme with intrinsic kinetics following the simple irreversible Michaelis-Menten rate law is considered above and throughout the remainder of this paper, the influence of diffusion when intrinsic kinetics follow more complex rate laws (e.g., substrate or product inhibition) has been analyzed elsewhere (Moo-Young and Kobayashi, 1972; Kerneves and Thomas, 1973), as have cases where more than a single reaction occurs (Goldman and Katchalski, 1971; Lawrence and Okay, 1973; Kerneves and Thomas, 1973), in which case effects on product selectivity may arise (Satterfield, 1970).

Electrostatic and pH effects. If the porous matrix is charged,
or if hydrogen ion is produced or consumed during reaction, then
pH within the matrix may differ from the pH of the external bulk
solution, and consequently observed kinetics can be modified
(Goldstein, et al., 1964; Goldman, et al., 1968; Laidler and
Sundaram, 1971; Bunow, 1974). When both substrate and porous
matrix are charged, additional effects, involving electrostatic
interaction between substrate and matrix, can be anticipated
(Goldstein et al., 1964; Hornby et al., 1968; Wharton, et al.,
1968; Shuler, et al., 1972, 1973; Hamilton, et al., 1973); however,
a rigorous analysis of this last problem appears difficult because,
for example, in a porous matrix large transverse concentration
gradients may develop as a result of electrostatic effects, and
the one-dimensional analysis of Equation (2) may become inadequate.

SOLUTIONS FOR EFFECTIVENESS FACTOR

Slab Geometry

Effectiveness factor concept. The effectiveness factor η is
defined as the observed rate of reaction in the membrane divided
by that which would obtain if the substrate concentration was
uniformly equal to its value at the membrane surface (Petersen,
1965; Gavalas, 1968; Satterfield, 1970; Aris, 1974):

$$v_{obs} = \eta \frac{V_m s_s}{K_m + s_s} \tag{6}$$

For Michaelis-Menten kinetics, the effectiveness factor is a
function solely of the modified Thiele modulus ϕ_m and the dimensionless Michaelis constant ν. The effectiveness factor concept
is useful because it allows prediction of the behavior of the
mass-transfer-influenced kinetics of an immobilized enzyme through
application of Equation (6).

Numerical solution. Problems mathematically equivalent to
the one of interest here, cast in terms of either Michaelis-Menten
kinetics or analogous Langmuir-Hinshelwood kinetics, were first
given limited treatment by Prater and Lago (1956), Blum and Jenden
(1957), and Chu and Hougen (1962), and subsequently more complete
numerical analysis by Roberts and Satterfield (1965), Krasuk and
Smith (1965), and Schneider and Mitschka (1965). The numerical
solution for η as a function of ϕ_m and ν is shown in graphical
form in Figure 1. Atkinson and Daoud (1968) fitted these numerical
results to a single empirical function by use of a least squares
procedure. We have found it convenient to employ the numerical
procedure of Roberts (1965) which involves selection of a value
for s at x = L followed by a marching finite-difference integration to solve for s_s and η.

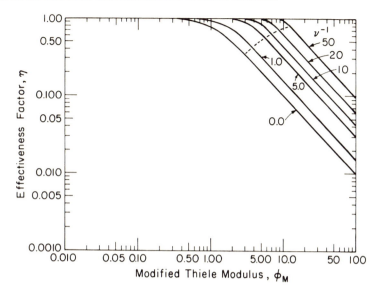

Figure 1:   Effectiveness factor η as a function of modified Thiele
            modulus ϕ_m and dimensionless Michaelis constant ν.
            Modified from Roberts and Satterfield (1965).

Asymptotic analytical solution.   An approximate, asymptotic
analytical solution is obtained if boundary condition II is replaced by:

IIa $s = 0$ at $x = L$ (7)

This corresponds to rapid reaction relative to diffusion (large ϕ_m)
in which case substrate does not penetrate far into the porous
matrix.   This notion has been discussed in depth by Bischoff (1965)
and Petersen (1965).   The asymptotic solution for effectiveness
factor (Roberts and Satterfield, 1965; Bischoff, 1965) is given by:

$$\eta \simeq \tilde{\eta} = \frac{\sqrt{2}}{\phi_m} \; (1 + \nu)\left[\nu^{-1} - \ln(1 + \nu^{-1})\right]^{1/2}$$
                                                                        (8)

where superscript tilde denotes that the solution is obtained with
boundary condition IIa.   Equation (8) is accurate within about
0.5% of the numerical solution anywhere below the dotted line
shown in Figure 1.

Zero- and first-order effectiveness factors. When ν^{-1} (dimensionless surface substrate concentration) is small compared to unity, one finds by expansion of Equation (8) that:

$$\tilde{\eta} \simeq \frac{1 + \frac{2}{3}\nu^{-1} + 0(\nu^{-2})}{\phi_m} \tag{9}$$

so that the asymptotic expression for a first-order reaction is recovered in the limit:

$$\lim_{\nu^{-1} \to 0} \tilde{\eta} = \lim_{\phi_m \to \infty} \eta_1 = \frac{1}{\phi_m} \tag{10}$$

where the first-order rate constant is given by V_m/K_m. Conversely, when ν^{-1} is much greater than unity, $\tilde{\eta}$ is given approximately by the expression for a zero-order reaction:

$$(\nu^{-1} \gg 1) \qquad \tilde{\eta} \simeq \eta_0 = \frac{\sqrt{2}}{\phi_m \nu^{1/2}} \tag{11}$$

which is valid for $\phi_m \nu^{1/2} > \sqrt{2}$ ($\eta_0 = 1$ for $\phi_m \nu^{1/2} \leq \sqrt{2}$). While zero- and first-order rate expressions have been used for convenience in several recent analyses (Van Duijn, et al., 1967; Goldman, et al., 1968; Sundaram, et al., 1970; Rony, 1971; Kasche, et al., 1971; Blaedel, et al., 1972; Lasch, 1972; Vieth, et al., 1973; Gondo, et al., 1973), it will be seen below that use of the full expression for $\tilde{\eta}$, Equation (8), yields more accuracy.

Other Geometries

The qualitative behavior of effectiveness factor plots is not very sensitive to geometry. In fact, when ϕ_m is sufficiently large that the reaction zone is confined to a relatively thin shell near the surface, the term accounting for curvature in spherical or cylindrical geometries can be dropped (Petersen, 1965; Gavalas, 1968), and the appropriate differential equation reduces to Equation (2). The problem analogous to the one treated here, but in a sphere with Langmuir-Hinshelwood kinetics has been solved numerically by Knudsen, et al. (1965) and Schneider and Mitschka (1966), and more recently for irreversible Michaelis-Menten kinetics by others (Horvath and Engasser, 1973; Fink, et al., 1973; Engasser and Horvath, 1973). The effectiveness factor for a sphere agrees closely with that for a slab (for all ϕ_m) if L in the definition of ϕ_m is replaced by V_p/S_p (particle volume/particle surface area), as suggested originally by Aris (1957).

EXTERNAL MASS TRANSFER RESISTANCE

In many situations, s_s is not directly measurable because external mass transfer influences can limit the observed rate of reaction (O'Neill, 1972). If an external concentration boundary layer is considered, it is easily shown (Hamilton, et al., 1974) that the bulk substrate and surface substrate concentrations are related by:

$$s_b = \frac{s_s}{K_p} + \frac{V_{obs} \, L}{k} \tag{12}$$

where s_b is the bulk concentration of substrate, k is the mass transfer coefficient, assumed constant over the surface of the porous medium, and the partition coefficient K_p is the equilibrium ratio at the membrane-solution interface of the substrate concentration within the membrane voids divided by the concentration in the external solution.

SHAPES OF TRADITIONAL PLOTS

General Behavior

Background. Graphical analysis of pseudo steady-state reaction rate data obtained with native enzymes in solution is commonly carried out on transformed coordinates such as Lineweaver-Burk, Eadie-Hofstee, or Woolf plots (Segel, 1968). These plots are useful in order to linearize data when kinetic rate laws for soluble enzymes follow certain prescribed patterns, for example, Michaelis-Menten kinetics. When enzymes are immobilized, however, Lineweaver-Burk, Eadie-Hofstee, or Woolf plots need not be linear, even if intrinsic kinetics follow the Michaelis-Menten rate law, because of effects arising from mass transfer. Such nonlinearity has been observed in several experimental studies (Van Duijn, et al., 1967; Lilly and Sharp, 1968; Kay and Lilly, 1970; Bunting and Laidler, 1972) and has been predicted theoretically (Moo-Young and Kobayashi, 1972; Kobayashi, et al., 1973; Engasser and Horvath, 1973; Hamilton et al., 1974). Nonlinear Lineweaver-Burk plots can also arise from other causes, such as polydispersity of intrinsic kinetic parameters (Kallen, et al., 1973), so that such nonlinearity is not a priori evidence of mass transfer effects.

Shapes of Lineweaver-Burk plots. Equation (6) can be rearranged into a form suitable for plotting, for example, on Lineweaver-Burk coordinates of reciprocal observed reaction rate versus reciprocal surface concentration:

$$\frac{1}{v_{obs}} = \frac{K_m}{\eta V_m} \frac{1}{s_s} + \frac{1}{\eta V_m} = \frac{1 + v}{\eta V_m} \tag{13}$$

If external mass transfer resistance is significant, s_s can be eliminated in favor of s_b through use of Equation (12). As pointed out by Engasser and Horvath (1973), similar expressions can be given for Eadie-Hofstee and Woolf plots. For purpose of illustration, we will give only a brief discussion of the shapes of Lineweaver-Burk plots; more detail is presented elsewhere (Hamilton, et al., 1974).

First consider cases for which the asymptotic solution $\eta \simeq \tilde{\eta}$ is valid. A combination of Equations (8) and (13) then yields:

$$\frac{1}{\tilde{v}_{obs}} = \frac{\phi_m}{\sqrt{2} \, V_m} \left[v^{-1} - \ln(1 + v^{-1}) \right]^{-1/2} \tag{14}$$

which, on Lineweaver-Burk coordinates, is clearly nonlinear. In fact, it can be shown (Hamilton, et al., 1974) that a Lineweaver-Burk plot for an immobilized enzyme whose intrinsic kinetics follow the Michaelis-Menten rate law (with V_m and K_m constant) is always concave to the $1/s_s$ axis in the region where Equation (14) is valid.

The limiting behavior of Lineweaver-Burk plots may be predicted from Equations (1), (9), (11), and (13). When v^{-1} is small with respect to unity,

$$\frac{1}{\tilde{v}_{obs}} \simeq \frac{\phi_m}{V_m} (v + \frac{1}{3}) \tag{15}$$

which yields a straight line with slope $\phi_m K_m / V_m$ and extrapolated intercept on the ordinate $\phi_m / 3 V_m$. If mass transfer effects were absent (Equation (1)), the slope would be K_m / V_m and the intercept $1/V_m$. With decreasing v^{-1}, one finds:

$$\lim_{v^{-1} \to 0} \frac{1}{\tilde{v}_{obs}} = \frac{\phi_m}{V_m} v \tag{16}$$

wherein the slope remains the same as in Equation (15) but the intercept is now zero, and the plot becomes indistinguishable from that expected for first-order kinetics at large ϕ_m.

At the other extreme, when $\nu^{-1} \gg 1$, the result expected for zero-order kinetics obtains:

$$\frac{1}{\tilde{v}_{obs}} \simeq \frac{\phi_m}{\sqrt{2}\ v_m}\ \nu^{1/2} \tag{17}$$

which is concave to the $1/s_s$ axis. It is clear that the predicted concave behavior of Lineweaver-Burk plots for immobilized enzymes is associated with the region where the asymptotic solution is valid. Conversely, it can be shown (Hamilton, et al., 1974) that near $\nu = 0$, where Equation (8) is no longer valid, there must be a region in which the Lineweaver-Burk plot is convex to the $1/s_s$ axis as the curve becomes asymptotic to the line which corresponds to Equation (1).

Numerical Examples

Several numerical examples (Hamilton, et al., 1974) are now given which illustrate the behavior predicted above. The example presented in Figures 2 and 3 displays all the characteristics that might be observed in a Lineweaver-Burk plot for an immobilized enzyme whose intrinsic kinetics follow the Michaelis-Menten rate law, with V_m and K_m constant. The relevant parameters were chosen so as to be typical of a physically realizable system. The results were generated using the numerical procedure of Roberts (1965) and the various analytical expressions described above.

Over virtually the entire range plotted in Figure 2, the numerical solution and asymptotic solution coincide, diffusion limitations are significant, and the effectiveness factor is much less than unity. For $\nu \gtrsim 1$, the curve converges asymptotically towards the line given by Equation (15), whereas agreement with the first-order approximation, Equation (16), is relatively poor.

The lower left-hand corner of Figure 2 is shown with an expanded scale in Figure 3. The curve is predominantly concave towards the abscissa with an inflection point where the numerical solution and asymptotic analytical solution diverge. The remainder of the curve at higher substrate concentration is convex as it asymptotically approaches the straight line which corresponds to the absence of diffusion limitations. The curve corresponding to zero-order kinetics shows only fair agreement with the numerical solution in the lower left-hand corner of the plot.

When external mass transfer effects are large, the behavior seen above becomes obscured. Figure 4 presents numerical examples

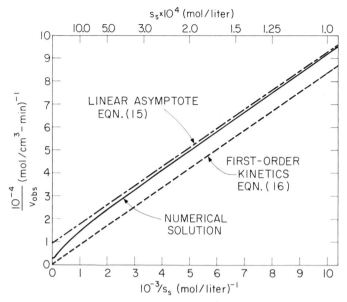

Figure 2:   Lineweaver-Burk plot calculated with the following
            parameters: $L = 5\times10^{-3}$ cm, $K_m = 3\times10^{-4}$M, D_{eff} =
            5×10^{-6} cm^2/sec, $V_m = 6\times10^{-6}$ mol product/(sec-cm^3 of support),
            $\phi_m = 10$. Maximum $\nu \simeq 3$. From Hamilton, et al. (1974).

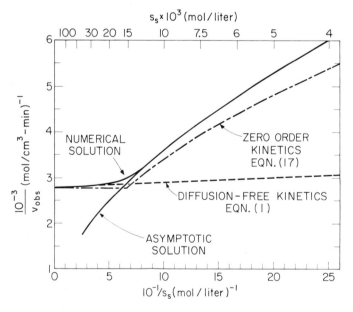

Figure 3:   Expanded plot of lower left-hand corner of Figure 2
            Maximum $\nu \simeq 0.075$. From Hamilton, et al. (1974).

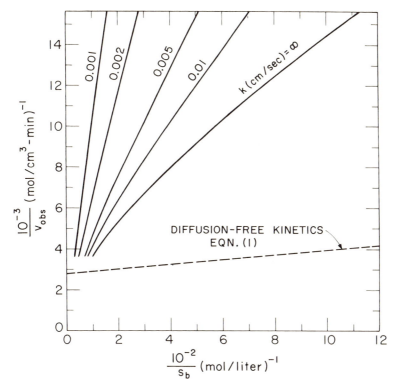

Figure 4: Effect of external mass transfer resistance on Lineweaver-Burk plots calculated with parameters used in Figures 2 and 3. From Hamilton, et al. (1974).

of Lineweaver-Burk plots calculated for a wide range of mass transfer coefficients with all other parameters the same as those used in Figures 2 and 3 and with K_p = 1.0. Through use of Equation (12), the abscissa is now expressed in terms of the bulk concentration, s_b instead of s_s. The curves are plotted over the region where the asymptotic analytical solution is valid, and therefore, the curve for infinite mass transfer coefficient is concave to the abscissa. As the external mass transfer resistance increases, the curvature decreases, and the curve for the lowest mass transfer coefficient is virtually linear. The

curves have a qualitative similarity to the experimental ob-
servations reported in Figure 9 of Lilly and Sharp (1968) where
decreasing agitation rate produced a more linear Lineweaver-Burk
plot.

THE APPARENT MICHAELIS CONSTANT

A recent recommendation (Sundaram, et al., 1972) suggests
reporting data in terms of $K_m(app)$ which is defined as the substrate concentration that gives a reaction velocity corresponding
to one-half $V_m(app)$. We assume here that $V_m(app)$ is the same as
V_m, an independent measure of which can be obtained, for example,
by making reaction rate measurements at sufficiently high substrate concentration (provided substrate inhibition is absent and
there are no substrate solubility limitations), and that the
substrate concentration with which calculations are made is the
surface substrate concentration obtained using Equation (12).
Equation (6) then can be written as:

$$\frac{v_{obs}}{V_m} = \frac{1}{2} = \frac{\eta}{1 + \dfrac{K_m}{K_m(app)}} \tag{18}$$

The value of ϕ_m for which Equation (18) is satisfied can be determined from the numerical and asymptotic solutions by sub-
stituting $K_m/K_m(app)$ for v(Hamilton, et al., 1974). The results
are shown in Figure 5. Given a measurement of $K_m(app)$, Figure 5
may be used to find the value of K_m which places the calculated
$K_m(app)/K_m$ and ϕ_m on the curve (or by implicit evaluation of the
asymptotic solution where applicable).

Iterative calculations can be eliminated by employing the
dimensionless modulus originally suggested by Wagner (1943):

$$\Phi_L = \frac{L^2 v_{obs}}{D_{eff} s_s} \tag{19}$$

which, for Michaelis-Menten kinetics, is also given by (Roberts
and Satterfield, 1965):

$$\Phi_L = \frac{\eta \phi_m^2}{1 + v^{-1}} \tag{20}$$

Figure 5 also presents a plot of $K_m(app)/K_m$ as a function of Φ_L.
The latter was calculated from Equation (20) with v^{-1} replaced
by $K_m(app)/K_m$, with $v_{obs} = V_m/2$, and with the corresponding values
of η and ϕ_m.

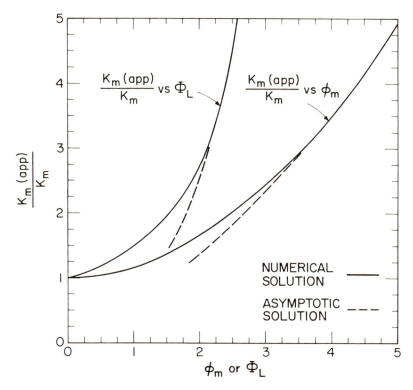

Figure 5: Dependence of $K_m(app)/K_m$ on ϕ_m or Φ_L. $K_m(app)$ is defined as s_s at which $v_{obs} = V_m/2$. On this plot, $\Phi_L = L^2 V_m/2 \, D_{eff} \, s_s$. (From Hamilton, et al. (1974).)

Figure 5 thus shows that it is possible to calculate the intrinsic K_m of an immobilized enzyme from a measured $K_m(app)$, as long as the assumptions cited above are met. Figure 5 also makes explicit the sensitive dependence of $K_m(app)$ on the mass transfer parameters (incorporated into ϕ_m or Φ_L) of the immobilized enzyme system.

DETERMINATION OF INTRINSIC KINETIC PARAMETERS

The technique for determining the values of the intrinsic kinetic constants V_m and K_m described in the preceding section is only one of several possible methods. Basically, there are three groups of methods for evaluating these intrinsic kinetic parameters. We list these three groups below, along with references which describe them in detail.

I. Variation of Substrate Surface Concentration

This group consists of methods which apply when a single type of immobilized enzyme module, of fixed shape, size, and loading, is used, but substrate concentration is varied. The K_m(app) method just described falls into this group, as do other procedures described by Bunting and Laidler (1972), Kobayashi and Laidler (1973), Engasser and Horvath (1973), and Hamilton, et al. (1974).

II. Variation of Characteristic Support Dimension or V_m

The "triangle method" of Weisz and Prater (1954), developed for analysis of the kinetics of heterogeneous catalysts, falls into this group. This method has been adapted for analysis of immobilized enzyme kinetics as described in a recent paper by Kobayashi and Laidler (1973). Engasser and Horvath (1973) have presented an additional method for evaluating the K_m of an immobilized enzyme by employing different size (or differently loaded) enzyme particles, as have Kobayashi and Laidler (1973).

III. Enzymatic Membrane Inserted Between Two Reservoirs

Selegny, et al. (1969) have described a procedure for evaluating the $\overline{K_m}$ of an enzyme immobilized in a membrane inserted between two reservoirs containing substrate at different concentrations. Somewhat similar work has been performed by others (Goldman, et al., 1968; Meyer et al., 1970; DeSimone and Kaplan, 1973). Workers using methods falling in this group are often interested in the use of membranes containing bound enzymes as models for more complex biological membranes.

Concluding Remarks

There does not yet appear to have been enough widespread experimental application of all the methods described above to conclude which of them in actual practice is most accurate and useful for determining the values of intrinsic kinetic parameters of immobilized enzymes. We assume that with the current extensive studies now underway in the immobilized enzyme area, enough consistent experimental data will appear in the next few years so that it will be possible to answer confidently the question

as to the extent to which intrinsic kinetic parameters of immobilized enzymes can be altered from values for native enzymes
free in solution, and how seriously such experimentally determined
alterations effect immobilized enzyme reactor design.

ACKNOWLEDGEMENTS

Supported in part by National Science Foundation Grant
Number GI 34284 and by a grant from the Camille and Henry Dreyfus
Foundation.

LITERATURE CITED

Aris, R., "On Shape Factors for Irregular Particles: I. The Steady
State Problem. Diffusion and Reaction," Chem. Eng. Sci., 6, 262
(1957).

Aris, R., "Mobility, Permeability, and the Pseudo-Steady-State
Hypothesis," Math. Biosci., 13, 1 (1972).

Aris, R., The Mathematical Theory of Diffusion and Reaction in
Permeable Catalysts, Oxford at the Clarendon Press, London (1974).

Atkinson, B., and S. Daoud, "The Analogy Between Microbiological
"Reactions" and Heterogeneous Catalysis," Trans. Instn. Chem.
Engrs., 46, T19 (1968).

Bischoff, K.B., "Effectiveness Factors for General Reaction Rate
Forms," AIChE J., 11, 351 (1965).

Blaedel, W.J., T.R. Kissel, and R.C. Boguslaski, "Kinetic Behavior
of Enzymes Immobilized in Artificial Membranes," Anal. Chem., 44,
2030 (1972).

Blum, J.J., and D.J. Jenden, "Rate Behavior and Concentration Profiles
in Geometrically Constrained Enzyme Systems," Arch. Biochem. Biophys.,
66, 316 (1957).

Bunow, B., "Enzyme Kinetics in Cells," J. Math. Biophys., in press
(1974).

Bunting, P.S., and K.J. Laidler, "Kinetic Studies on Solid-
Supported β-Galactosidase," Biochemistry, 11, 4477 (1972).

Chu, C., and O.A. Hougen, "The Effect of Adsorption on the
Effectiveness Factor of Catalyst Pellets," Chem. Eng. Sci., 17,
167 (1962).

Cleland, W.W., "The Statistical Analysis of Enzyme Kinetic Data," Adv. Enzymol. (edited by F.F. Nord), Vol. 29, Interscience, New York, page 1 (1967).

Coleman, M.H., "Graphical Analysis of Enzyme Kinetic Data," Nature, 205, 798 (1965).

Colton, C.K., K.A. Smith, E.W. Merrill, and P.C. Farrell, "Permeability Studies with Cellulosic Membranes," J. Biomed. Mater. Res., 5, 459 (1971).

Crank, J., The Mathematics of Diffusion, Oxford at Clarendon Press, London (1956).

Desimone, J.A., and S.R. Caplan, "Symmetry and the Stationary State Behavior of Enzyme Membranes," J. Theor. Biol., 39, 523 (1973).

Dixon, M., and E.C. Webb, Enzymes, Academic Press, New York (1964).

Dowd, J.E., and D.S. Riggs, "A Comparison of Estimates of Michaelis-Menten Kinetic Constants from Various Linear Transformations," J. Bio. Chem., 240, 863 (1965).

Engasser, J., and C. Horvath, "Effect of Internal Diffusion in Heterogeneous Enzyme Systems: Evaluation of True Kinetic Parameters and Substrate Diffusivity," J. Theor. Biol., 42, 137 (1973).

Fink, D.J., T. Na, and J.S. Schultz, "Effectiveness Factor Calculations for Immobilized Enzyme Catalysis," Biotech. Bioeng., 15, 879 (1973).

Gavalas, G.R., Nonlinear Differential Equations of Chemically Reacting Systems, Springer-Verlag, New York (1968).

Goldman, R.O., and E. Katchalski, "Kinetic Behavior of a Two-Enzyme Membrane Carrying Out a Consecutive Set of Reactions," J. Theor. Biol., 32, 243 (1971).

Goldman, R., O. Kedem, and E. Katchalski, "Papain-Collodion Membranes. II. Analysis of the Kinetic Behavior of Enzymes Immobilized in Artificial Membranes," Biochemistry, 7, 4518 (1968).

Goldstein, L., Y. Levin, and E. Katchalski, "A Water-insoluble Polyanionic Derivative of Trypsin. II. Effect of the Polyelectrolyte Carrier on the Kinetic Behavior of the Bound Trypsin," Biochemistry, 3, 1913 (1964).

Gondo, S., T. Sato, and K. Kusunoki, "Note on the Lineweaver-Burk Plots for the Immobilized Enzyme Particle," Chem. Eng. Sci., 28, 1773 (1973).

Hamilton, B.K., L.J. Stockmeyer, and C.K. Colton, "Comments on Diffusive and Electrostatic Effects with Immobilized Enzymes," J. Theor. Biol., 41, 547 (1973).

Hamilton, B.K., C.R. Gardner, and C.K. Colton, "Effect of Diffusional Limitations on Lineweaver-Burk Plots for Immobilized Enzymes," AIChE J., in press (1974).

Heineken, F.G., H.M.,Tsuchiya, and R. Aris, "On the Mathematical Status of the Pseudo-steady State Hypothesis of Biochemical Kinetics," Math. Biosci., 1, 95 (1967).

Hornby, W.E., M.D. Lilly, and E.M. Crook, "Some Changes in the Reactivity of Enzymes Resulting from their Chemical Attachment to Water-Insoluble Derivatives of Cellulose," Biochem. J., 107, 669 (1968).

Horvath, C. and J.M. Engasser, "Pellicular Heterogeneous Catalysts. A Theoretical Study of the Advantages of Shell Structured Immobilized Enzyme Particles," Ind. Eng. Chem. Fund., 12, 229 (1973).

Kallen, R.G., T. Newirth, and M. Diegelman, "Consecutive Reactions with Immobilized Enzymes," presented at 66th Annual AIChE Meeting, Philadelphia, Pa., Nov. 11-15, 1973.

Kasche, V., H. Lundqvist, R. Bergman, and R. Axen, "A Theoretical Model Describing Steady-State Catalysis by Enzymes Immobilized in Spherical Gel Particles. Experimental Study of α-Chymotrypsin-Sepharose," Biochem. Biophys. Res. Commun., 45, 615 (1971).

Kay, G., and M.D. Lilly, "The Chemical Attachment of Chymotrypsin to Water-Insoluble Polymers Using 2-Amino-4, 6-dichloro-5-triazine," Biochem. Biophys. Acta, 198, 276 (1970).

Kernevez, J.P., and D. Thomas, "Numerical Analysis of Immobilized Enzyme Systems," Rapport de Recherche No. 28, Institute de Recherche d'Informatique et d'Automatique, Domaine de Voluceau, Rocquencort, Le Chesnay, France, Sept. 1973.

Kobayashi, T., G. Van Dedem, and M. Moo-Young, "Oxygen Transfer into Mycelial Pellets," Biotech. Bioeng., 15, 27 (1973).

Kobayashi, T., and K.J. Laidler, "Kinetic Analysis for Sodid-Supported Enzymes," Biochem. Biophys. Acta, 302, 1 (1973).

Knudsen, C.W., G.W. Roberts, and C.N. Satterfield, "Effect of Geometry on Catalyst Effectiveness Factor: Langmuir-Hinshelwood Kinetics," Ind. Eng. Chem. Fund., 5, 325 (1966).

Krasuk, J.H., and J.M. Smith, "Effectiveness Factors with Surface Diffusion," Ind. Eng. Chem. Fund., 4, 102 (1965).

Laidler, K.J., and P.V. Sundaram, "The Kinetics of Supported Enzyme Systems," in Chemistry of the Cell Interface (edited by H.D. Brown), Academic Press, New York (1971).

Lasch, J., "Theoretical Analysis of the Kinetics of Enzymes Immobilized in Spherical Pellets," in Analysis and Simulation of Biochemical Systems, edited by H.C. Henker and B. Hess, American Elsevier, New York, page 295 (1972).

Lawrence, R.L., and V. Okay, "Diffusion and Reaction in a Double Enzyme Supported Catalyst," Biotech. Bioeng., 15, 217 (1973).

Lilly, M.D., and A.K. Sharp, "The Kinetics of Enzymes Attached to Water-Insoluble Polymers," The Chemical Engineer, Jan/Feb, CE12 (1968).

Marsh, D.R., Y.Y. Lee, and G.T. Tsao, "Immobilized Glucoamylase on Porous Glass," Biotech. Bioeng., 15, 483 (1973).

Meyer, J., F. Sauer, and D. Woermann, "Study of a First Order Diffusion Controlled Chemical Reaction Occuring Inside Catalytically Active Membranes," Berichte der Bunsen-Gesellschaft, 74, 245 (1970).

Miyamoto, K., T. Fujii, N. Tamaoki, M. Okazaki, and Y. Miura, "Intraparticle Diffusion in the Reaction Catalyzed by Immobilized Glucoamylase," J. Ferment. Technol., 51, 566 (1973).

Moo-Young, M., and T. Kobayashi, "Effectiveness Factors for Immobilized-Enzyme Reactions," Can. J. Chem. Eng., 50, 162 (1972).

O'Neill, S.P., "External Diffusional Resistance in Immobilized Enzyme Catalysis," Biotech. Bioeng., 14, 675 (1972).

Petersen, E.E., Chemical Reaction Analysis, Prentice-Hall, Englewood Cliffs (1965).

Plowman, K.M., Enzyme Kinetics, McGraw-Hill, New York (1972).

Prater, C.D., and R.M. Lago, "The Kinetics of the Cracking of Cumene by Silica-Alumina Catalysts," in Adv. in Cat. (edited by W.G. Frandenburg and V.I. Domarensky), Vol. 8, Academic Press, New York (1956).

Roberts, G.W., "I. Kinetics of Thiophene Hydrogenolysis. II. Effectiveness Factors for Porous Catalysts," Sc.D. Thesis, Mass. Inst. of Technology, Cambridge (1965).

Roberts, G.W., and C.N. Satterfield, "Effectiveness Factor for Porous Catalysts," _Ind. Eng. Chem. Fund._, 4, 289 (1965).

Rony, P.R., "Multiphase Catalysis. II. Hollow Fiber Catalysts," _Biotech. Bioeng._, 13, 431 (1971).

Roughton, F.J.W., "Diffusion and Chemical Reaction Velocity as Joint Factors in Determining the Rate of Uptake of Oxygen and Carbon Monoxide by the Red Blood Corpuscle," _Proc. Roy. Soc. B_, 111, 1 (1932).

Roughton, F.J.W., "Diffusion and Chemical Reaction Velocity in Cylindrical and Spherical Systems of Physiological Interest," _Proc. Roy. Soc. B_, 140, 203 (1952).

Roughton, F.J.W., "Diffusion and Simultaneous Chemical Reaction Velocity in Haemoglobin Solutions and Red Cell Suspensions," _Prog. Biophys. Mol. Biol._, 9, 55 (1959).

Rovito, B.J., and J.R. Kittrell, "Film and Pore Diffusion Studies with Immobilized Glucose Oxidase," _Biotech. Bioeng._, 15, 143 (1973).

Satterfield, C.N., _Mass Transfer in Heterogeneous Catalysis_, MIT Press, Cambridge (1970).

Schneider, P., and P. Mitschka, "Effect of Internal Diffusion on Catalytic Reaction," _Collection Czechoslov. Chem. Commun._, 30, 146 (1965).

Schneider, P., and P. Mitschka, "Effect of Internal Diffusion on Catalytic Reactions. III. Effect of Particle Shape on Reaction with Langmuir-Hinshelwood Type of Kinetics," _Collection Czechoslov. Chem. Commun._, 31, 1205 (1966).

Segel, I.H., _Biochemical Calculations_, Wiley, New York (1968).

Selegny, E., G. Brown, J. Geffroy, and D. Thomas, "Méthode de Détermination de K_M Réel d'un Enzyme par le Régime Stationnaire d'une Membrane en Activité Enzymatique," _J. Chem. Phys. et Physico-Chimie Bio._, 66, 391 (1969).

Shuler, M.L., H.M. Tsuchiya, and R. Aris, "Diffusive and Electro-static Effects with Insolubilized Enzymes," _J. Theor. Biol._, 35, 67 (1972).

Shuler, M.L., H.M. Tsuchiya, and R. Aris, "Diffusive and Electro-static Effects with Insolubilized Enzymes Subject to Substrate Inhibition," _J. Theor. Biol._, 41. 347 (1973).

Sundaram, P.V., A. Tweedale, and K.J. Laidler, "Kinetic Laws for Solid-Supported Enzymes," Can. Jour. Chem., 48, 1498 (1970).

Sundaram, P.V., E.K. Pye, T.M.S. Chang, V.H. Edwards, E.A. Humphrey, N.O. Kaplan, E. Katchalski, Y. Levin, M.D. Lilly, G. Manecke, K. Mosbach, A. Patchornik, J. Porath, H.H. Weetall, and L.B. Wingard, Jr., "Recommendations for Standardization of Nomenclature in Enzyme Technology," Enzyme Engineering (edited by L.B. Wingard, Jr.), Wiley-Interscience, page 15 (1972).

Thomas, D., G. Brown, and E. Selegny, "Monoenzymatic Model Membranes: Diffusion-Reaction Kinetics and Phenomena," Biochimie, 54, 229 (1972).

Van Duijn, P., E. Pascoe, and M. Van der Ploeg, "Theoretical and Experimental Aspects of Enzyme Determination in a Cytochemical Model System of Polyacrylamide Films Containing Alkaline Phosphatase," J. Hist. Cytochem., 15, 631 (1967).

Vieth, W.R., A.K. Mendiratta, A.O. Mogensen, R. Saini, and K. Venkatasubramanian, "Mass Transfer and Biochemical Reaction in Enzyme Membrane Reactor Systems. I. Single Enzyme Reactions," Chem. Eng. Sci., 28, 1013 (1973).

Wagner, C., "Über das Zusammenwinker von Strömung, Diffusion und chemischer Reaktion bei der heterogenen Katalyse," Z. Phys. Chem., 193, 1 (1943).

Weisz, P.B., and C.D. Prater, "Interpretation of Measurements in Experimental Catalysts," Adv. Catalysis, 6, 143 (1954).

Wharton, C.W., E.M. Crook, and K. Brocklehurst, "The Nature of the Perturbation of the Michaelis Constant of the Bromelain-Catalyzed Hydrolysis of α-N-Benzoyl-L-Arginine Ethyl Ester Consequent upon Attachment of Bromelain to O-(Carboxymethyl)-Cellulose," Eur. J. Biochem., 6, 572 (1968).

A COMPARISON OF PROPOSED METHODS FOR THE IN VITRO SYNTHESIS OF EDIBLE CARBOHYDRATES

J. L. Adams, J. Billingham and J. Shapira

Ames Research Center, NASA, Moffett Field, California,

and Stanford University, Stanford, California

INTRODUCTION AND SUMMARY

General

This report is the result of an eleven week systems design project sponsored jointly by the National Aeronautics and Space Administration and the American Society of Engineering Education. The group consisted of eighteen faculty members from across the nation, representing a number of disciplines. Stanford University and Ames Research Center were the host institutions for the project. The purpose was both to give the participants experience in systems engineering and to produce a useful study.

The project involved applying a systems approach to study nonagricultural carbohydrate production on a large scale. The project grew out of work done at the Ames Research Laboratory in the general area of synthesizing food during long-duration manned space missions. Since approximately 80% by weight of a person's diet can be carbohydrate, and since carbon dioxide and water are available as by products from the metabolic cycle, the Ames work was focused toward the production of carbohydrate from carbon dioxide and water. Once into this work, it was impossible to avoid considering the possibilities of using such a synthetic process to provide food on a large scale here on earth.

Even the most optimistic forecasts of world population growth indicate a severe challenge to the ability of agriculture to meet future food needs. Figure 1 shows the classical long range trend diagram for human population growth. Malnutrition and

225

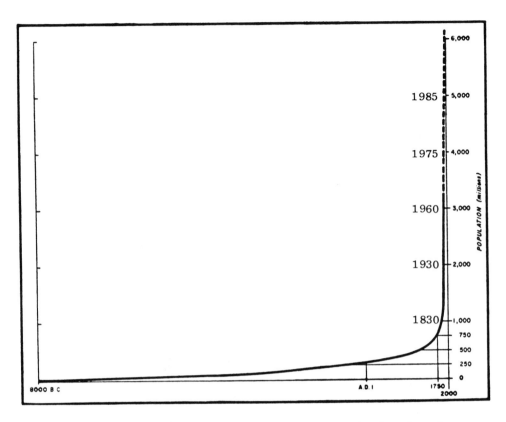

Figure 1. Long-Range Trend of World Population Growth

under-nutrition continue to be serious problems in many parts of
the world. Even projections which estimate a world population on
the order of 7 billion by the year 2000, fail to depict the near
term plight of the developing countries where the food needs in 1985
will be 2-1/2 times that of 1962. Figures 2 and 3 give differing
estimates of anticipated population growth.

Although there is great potential for increasing food production in the so-called "Green Revolution," it must be recognized that
there are limits to the earth's land, water, and mineral resources,
and hence to the ultimate capacity of agriculture systems. In Asia,
where the population problem is currently the most severe, essentially all of the arable land is already in use. Better control of
water supplies implying drainage as well as irrigation, increased
fertilizer use, and modern farming techniques all involve high
capital costs.

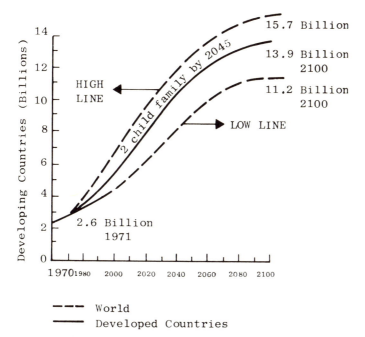

Figure 2. "Gradual Approach to Zero Population Growth"

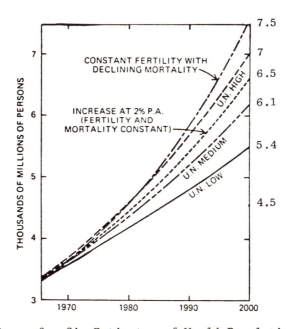

Figure 3. Six Estimates of World Population

SYNTHETIC FOODS

History

An alternative to agricultural production of food is to
utilize chemical and/or enzymatic processes to produce materials
which could become a significant proportion of the diet. The ra-
tionale is that food is a complex mixture of compounds, some of
which might be amenable to efficient synthesis. Thus, the possibility exists of assembling "food factories" which would convert inexpensive starting materials into safe and acceptable food components.

The advantages of such a food production system are many. These
include high efficiency of land utilization, flexibility of location,
independence from climate or type of soil, minimal water requirements,
lack of agricultural waste and no environmental pollution from fertilizers or insecticides. This approach of producing food non-
agriculturally has some precedent.

Carbohydrates

Considering the difficulties in the physicochemical synthesis of
edible fat and protein, it is not surprising that the major recent
effort has been directed toward carbohydrate. One potential method
that has received considerable attention has been the formose reaction. In this reaction, certain alkalis and other substances catalyze the self condensation of formaldehyde to produce a mixture of
monosaccharides. The resulting mixture has not been well character-
ized in most of the reported studies dealing with changes in the
nature of the product with changes in catalyst or conditions. Relatively recent studies using gas-liquid chromatography have shown
that the product is usually very complex with relatively small
amounts of metabolizable carbohydrate. Efforts to obtain simpler
mixtures have not been very successful. It should be noted that
the formose reaction gives rise to both the D- and L-isomer of any
given product and therefore it should be expected that even if it
were possible to control the reaction to produce only hexoses, maxi-
mally only one-half the product could be metabolized.

There are few reports in the literature concerning the nutritive
qualities of formose sugars. One would expect the results to be
quite variable depending upon the catalysts and conditions used in
their preparation. Mixed microorganisms can utilize a considerable
proportion of the mixture for growth. When fed to rats as 40 percent
of the diet, formose sugars caused diarrhea and death. At lower
levels, the animals survived but suffered from diarrhea. As yet, the
reason for this effect is not known.

Other carbohydrate-like materials such as glycerol have been considered as potential nutrients. Glycerol is normally present in the diet as a component of lipids and in the free form. It can be tolerated in relatively large amounts in the human diet without obvious detrimental effect.

THE STUDY

Consideration of population trends, the political, economic, and legal aspects associated with food production and consumption, and recent advances in enzyme chemistry, led the group to consider three chemical pathways which would lead to the production of glucose and starch. The first is a reasonably direct process which digests cellulose into glucose. The second is a comparatively simple enzyme-mediated process which converts a petrochemical into carbohydrates. The third is a process which mimics some aspects of photosynthesis, using hydrogen to reduce carbon dioxide to sugar in approximately 17 steps.

This paper outlines each of these three processes and points out advantages, disadvantages, and critical problems involved with each. The "cellulose" and "fossil fuel" processes were carried into fairly detailed designs of plants with 100 tons/day capacity. The "CO_2 fixation" process was not carried to detailed plant design, but was given a thorough feasibility study.

Cellulose Process

The production of food from cellulose has a reasonably long history. Acid hydrolysis of wood chips was carried out on an industrial scale both in the U.S. and in Germany during the first half of this century. However, the use of enzymatic degradation has been a comparitively recent development. Two groups, one at Louisiana State University and one at University of California at Berkeley, have studied similar processes with similar results in terms of cost data.

The cellulose process utilizes the extracellular enzymes of the mold _Trichoderma viride_ to digest inedible cellulose polymers to glucose. In the chemical plant designed by this group, bagasse (sugar cane waste) is used as raw material. After mechanical disruption by chopping and milling, alkali treatment serves to loosen up the structure sufficiently to allow attack by the cellulases from the fungus. (It also dissolves off the hemicellulose impurities contained in the raw material.) The pretreated material is then sent to a reactor along with the enzyme containing exudate from the _Trichoderma_. (The organism is grown in a fermenter in parallel with the reactor and uses the raw bagasse as an energy source in the

culture medium.) After hydrolysis in the reactor, which produces
a 10 to 15% glucose solution, the contents are filtered, allowing the
product glucose to pass, but retaining the enzyme and unused cellulose for recycling and reuse.

Starch, rather than glucose, offers at least three attractive
possibilities for improving food supplies: (1) as a basic component
in a prepared food, (2) as an extender for conventional food components, and (3) as a basic component in animal feeds. The first is
the most appealing from the standpoint of increasing the quantity of
foods meeting basic nutritional needs. A high starch product
widely used in Japan is "instant Ramen," a noodle packed in individual
servings which requires only the addition of boiling water for preparation. Such products might well find wide acceptance in many
regions of the world and could be easily fortified and flavored to
suit regional tastes. Starch is already used with flour in baked
goods in many parts of the world and commercial recipe formulations
could dramatically increase that use. The third use, as animal feed,
offers the promise of improving diets through meat protein, most
likely chicken. Chicken feeds are high in energy, and the energy
component represents close to one-half of feed costs. Consumer
acceptance of starch as a food would no longer present a problem, and
availability of food grains for human use would be increased.

The glucose produced in this process can be converted to starch
by three additional steps using microbial enzymes. The first two
steps add a phosphate to glucose. Due to the peculiarities of available enzymes this must be done by adding it first to one carbon atom
in the glucose molecule and then transferring it. The source of the
phosphate is ATP, a well known but expensive biochemical, and two
enzymes are used, hexokinase from yeast and phosphoglucomutase from
Escherichia coli. The third step causes the phosphorylated glucose
to polymerize, giving starch. (Figure 4.)

This segment of the process presented several challenges. There
are few design precedents on which to base estimates and to delineate
problems, as enzymatic conversions on an industrial scale have been
restricted to very small quantities of material (as in drug production) or have involved the use of whole organisms in fermentations
(as in wine production).

First, the large scale initial production of three relatively
purified enzymes is required, and although yeast and E. coli (the
sources) are easily grown, about 50 tons for a 100 ton/day plant are
needed. The production facility must also necessarily provide for
replenishment of these enzymes as they deteriorate.

Another concern is the requirement that adenosine triphosphate
(ATP) be constantly regenerated. Because of the cost of ATP, the

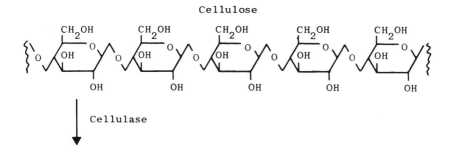

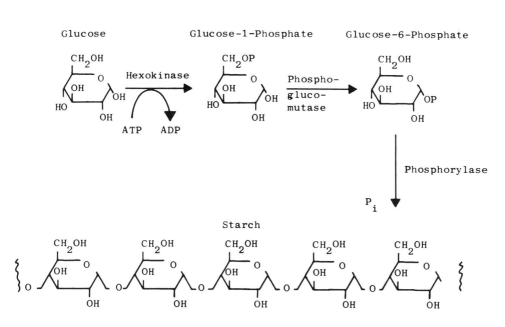

Figure 4.   Chemical Pathway-Cellulose to Starch

breakdown products of ATP which are sloughed off at different points must be recombined. In order to accomplish this, a satellite regenerator is proposed (Figure 5). In this regenerator, the breakdown products (adenosine diphosphate (ADP) and inorganic phosphate (P_i)] are mixed in the presence of potassium cyanate and an enzyme from E. coli which condenses the intermediately formed carbamyl phosphate with ADP to form ATP. To complete the cycle, the potassium cyanate is also regenerated from the products of this reaction.

A third problem was the design of separation systems which would discriminate on a large scale between such similarly charged chemicals as ADP and glucose-6-phosphate. Ion exchange chromatography is planned because the alternative method, selective membranes, requires excessive areas and pressures. This involves considerable scaling up of what is ordinarily a laboratory procedure. It is also necessary to glean the final starch product away from its smaller precursors and the enzymes that make it. The problem of precipitating starch and allowing the recovery of the soluble enzymes, as well as recycling the phosphate and unreacted sugar-phosphate, involves a complex regime of centrifugation, ion exchange adsorption and enzyme stabilization. (Figure 6.) The complete cellulose to glucose flow diagram is shown in Figure 7.

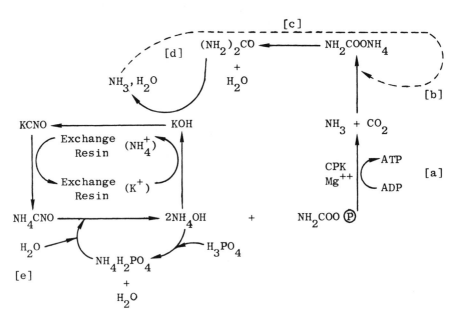

Figure 5. ATP Regeneration-Cyanate Process

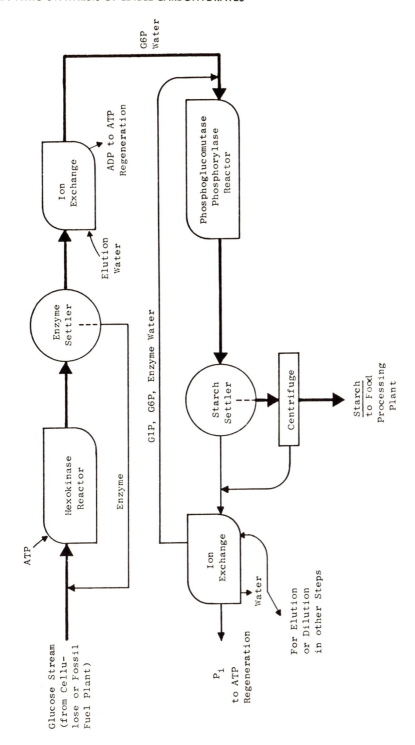

Figure 6. Glucose to Starch Flow Diagram

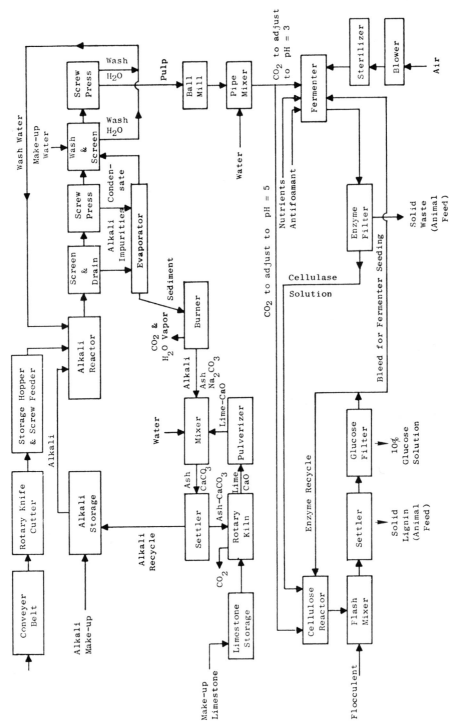

Figure 7. Cellulose to Glucose Flow Diagram

The cost of construction of a plant to produce 100 tons of glucose from cellulose per day is estimated to be about $3.5 million. At operating and material costs based on present day prices, the glucose could be produced at about 4.5¢ per pound. Cost estimates for the conversion of glucose to starch are 5.8¢/lb. This gives a total cost per pound of starch produced from cellulose as 10.3¢/lb. (This is in the range of current carbohydrate costs which range from 2¢ to 18¢ per pound of metabolizable starch.) To the extent that glucose could be tapped off and used as a food supplement, the price would be even more competitive.

A sugar cane mill processing 4000 tons of cane per day for a four month crushing season would provide enough bagasse to operate a 100 ton per day plant the year round. Bagasse can be easily handled and stored to retain its value as a raw material.

Fossil Fuel Process

The fossil fuel process was looked upon more as a design challenge than as an immediately applicable solution to the world's food problems.

The sequence uses glycidaldehyde, an intermediate in the industrial conversion of petroleum to glycerol, as starting material. The glycidaldehyde is distilled into aqueous solution in contact with an acid resin to convert it to a mixture of glyceraldehyde and dihydroxyacetone, both 3-carbon sugars. An enzyme, triokinase, then phosphorylates these small sugars (trioses) so that they can be condensed using the enzyme aldolase to fructose diphosphate (FDP) (a phosphorylated 6-carbon sugar). The FDP is then hydrolyzed to fructose (a component of invert sugar and honey) which could, if desired, be tapped off as a product. Since fructose is 1.7 times as sweet as sugar, it is a valuable food chemical.

A portion of the fructose is then converted to glucose using glucose isomerase (an enzyme which has been used on an industrial scale by the Japanese to carry out the reverse reaction in order to sweeten syrup). The glucose is then converted to starch in a three step process identical with that used in the cellulose route. The glycidaldehyde to glucose scheme is shown in Figure 8, and the glycidaldehyde to starch pathway in Figure 9.

In the total process, the large scale initial production of eight enzymes along with provision for their replenishment is required. In general, microbial sources are used--three of the seven prepared from yeast, three from E. coli, one from Streptomyces albus, and one from a Bacillus subtilis mutant. The quantities of starting materials, in terms of dry cell weight, range from 48 lbs. (to isolate triose phosphate isomerase) to 70 tons (to isolate the triokinase).

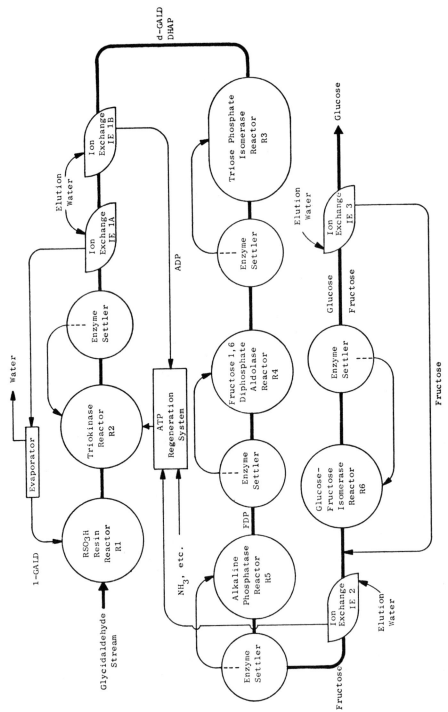

Figure 8. Chemical Reactions–Glycidaldehyde → Glucose

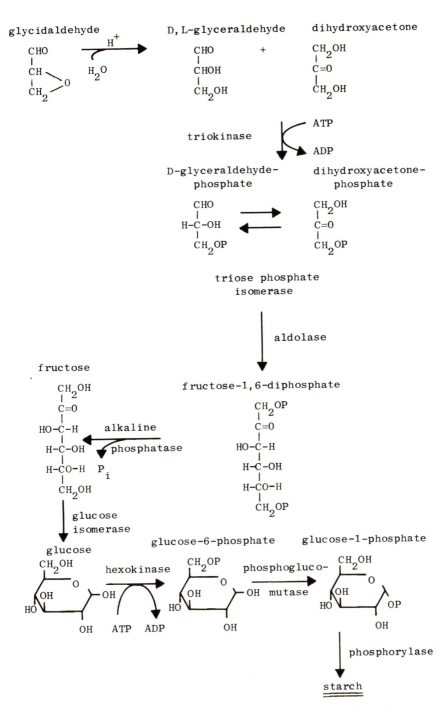

Figure 9. Chemical Pathway-Glycidaldehyde to Starch

The design calls for running the process as far as glucose-6-phosphate with stirred tank reactors using the enzymes on solid supports so that they can be filtered out. Four of the six enzymes in this portion of the sequence have been immobilized on various particle backings. It is assumed that attempts to apply these techniques to the remaining two would be successful.

The projected production cost of glucose from this process would be 10.1¢ per pound exclusive of the raw material, and carrying this to starch in the same facility would add another 5¢. The total cost of 15.1¢ per lb would then be in the competitive range if it were not for the cost of the glycidaldehyde, which is estimated at 17¢/lb. Since the precursor of glycidaldehyde (propylene) is quite cheap (corresponding to 1.5¢ per lb of carbohydrate) a breakthrough in this industrial process could make the process practical.

CO$_2$ Fixation Process

The CO$_2$ fixation process (Figure 10) is by its very nature an order of magnitude more complex than the fossil fuel process. There are 16 steps in the route from CO$_2$ to starch, catalyzed by 13 different enzymes, but the additional number of steps is the least problem.

The pathway is cyclic. In plants, a single CO$_2$ is attached to a 5-carbon recipient to give a 6-carbon compound. However, this does not work unless the 5-carbon recipient can be regenerated. Green plants do this by restructuring 5/6 of these 6-carbon products into 5-carbon recipients. This means the process cannot be carried out in a straightforward way, putting in reactants at the beginning, and getting out products at the end. It implies, rather, a controlled separation of 1/6 of the product formed at some point.

Regeneration processes are also escalated; in the formulation of one glucose, 18 ATP molecules are required, rather than 2 as in the fossil fuel sequence, or 0 as in the cellulose process. ATP is still needed in the glucose-to-starch sequence, and a whole new problem arises due to the oxidation state of CO$_2$, the starting material. Reducing power needs to be supplied in the form of NADPH, an unstable biochemical which is oxidized in the process to nicotinamide adenine dinucleotide phosphate (NADP$^+$), and must, of course, be re-reduced. This re-reduction implies the use of an external material rather than recyclable intermediates. The possibility of using alcohol dehydrogenase to catalyze its reduction by ethyl alcohol was explored but the enzyme-mediated use of H$_2$ to reduce ferridoxin (a chemical serving this function in the natural photosynthetic process) was preferred. There is currently no good experimental precedent on which to base this.

Number	Reaction
1	$6(\text{RuDP} + CO_2 + H_2O) \rightleftarrows 2\ 3\text{-PGA} + 2H^+)$
2	$12(3\text{-PGA} + \text{ATP} \rightleftarrows \text{P-3PGA} + \text{ADP})$
3	$12(\text{P-3PGA} + \text{NADPH} + H^+ \rightleftarrows \text{GALD3P} + \text{NADP}^+ + \text{HPO}_4^=)$
4	$5(\text{GALD3P} \rightleftarrows \text{DHAP})$
5	$3(\text{GALD3P} + \text{DHAP} \rightleftarrows \text{FDP})$
6	$3(\text{FDP} + H_2O) \rightleftarrows \text{F6P} + H$
7	$2(\text{F6P} + \text{GALD3P} \rightleftarrows \text{E4P} + \text{Xu5P})$
8	$2(\text{E4P} + \text{DHAP} \rightleftarrows \text{SDP})$
9	$2(\text{SDP} + H_2O \rightleftarrows \text{S7P} + \text{HPO}_4^=)$
10	$2(\text{S7P} + \text{GALD3P} \rightleftarrows \text{Xu5P} + \text{R5P})$
11	$2(\text{R5P} \rightleftarrows \text{Ru5P})$
12	$4(\text{Xu5P} \rightleftarrows \text{Ru5P})$
13	$6(\text{Ru5P} + \text{ATP} \rightleftarrows \text{RuDP} + \text{ADP} + H^+)$
14	$\text{F6P} \rightleftarrows \text{G6P}$
15	$\text{G6P} \rightleftarrows \text{G1P}$
16	$\text{G1P} \rightleftarrows \text{starch}\ (C_6H_5O_{10}) + \text{HPO}_4^=$
Net reaction:	$6CO_2 + 12\text{NADPH} + 18\text{ATP} + 11H_2O \rightarrow$
	$\text{Starch} + 12\text{NADP}^+ + 18\text{ADP} + 18\text{HPO}_4^= + 6H^+$

Figure 10.   Reactions of the Carbon Dioxide Fixation Process

Enzyme production is also a source of considerable concern. Three of the necessary enzymes have so far been prepared only in crude form. Only four of the 12 enzymes in the sequence from CO_2 to glucose have been immobilized on solid supports. Quantities required also appear to be high. There would be obvious advantage in obtaining all the enzymes from one source, such as spinach, but even operating at the maximum efficiency of the green plant itself, 500 tons of spinach would be required. If purified enzymes are needed, the amounts increase to 40,000 tons or so of source material.

When these difficulties are quantified, the increments added to the cost of product starch are extremely high. The make-up quantities of NADPH and ATP alone are excessive even at very low attrition rates. In view of this, no attempt was made to design a 100 ton/day factory in detail, but a preliminary cost estimate of 77¢ per pound has been made.

CO_2 FIXATION PROCESS-DESIGN CONSIDERATIONS--CO_2 TO GLUCOSE

Single Reactor Scheme

The analogy to the single reactor scheme is the chloroplast, which carries out all the reactions proposed plus many others. In addition to converting CO_2 to starch by the pathway shown, it regulates concentrations of reactants and products, converts water to hydrogen and oxygen, phosphorylates ADP to make ATP using energy from sunlight, and synthesizes many of the enzymes involved. In the single reactor scheme proposed here, enzymes would catalyze the reactions converting CO_2 to starch. The entering materials would be hydrogen (to regenerate NADPH from $NADP^+$) and carbamyl phosphate (to regenerate ATP and provide the CO_2). Starch, $MgNH_4PO_4 \cdot 6H_2O$ (magnesium ammonium phosphate) and CO_2 would be taken out. These substances are easy to separate since starch and $MgNH_4PO_4 \cdot 6H_2O$ are solids, while carbon dioxide is gaseous.

The advantage of this setup is that it is mechanically simple and avoids many difficult separations. It obviously has the precedent of the green plant cell itself. However, the question of localization of cyclic enzymes is still open. There is as yet no definite evidence that the enzymes in the chloroplast are organized but the high concentrations of cyclic enzymes in the stroma region may imply that their movement is restricted and that there is some structure.

There are a number of disadvantages to a single reactor. The multiplicity of intermediates leads to the possibility of lower rates due to more inhibitors (though there may be more activators also). It would also be extremely difficult to design, without a major

experimental program, since very little is known at present of the
reaction kinetics. There is also a problem in controlling each of
the reactions as their individual enzymes decay with time. Since the
many enzymes involved undoubtedly decay at different rates, in order
to keep all at an adequate activity level, those that decay quickly
will have to be separated from those that decay more slowly. Additionally, the question of nonspecificity (the case where one enzyme
catalyzes more than one reaction) is still unresolved. There are
two enzymes in the process, transketolase and aldolase, which catalyze formation of extraneous products from intermediates present in
the reactor that are part of the process. These products, if
accumulated, would serve to drain intermediates from the cycle. They
do not accumulate in a normally functioning cell and it is unknown
how much they would accumulate in the proposed reactor.

It is as yet difficult to carry out the reactions of the CO_2
fixation pathway by using enzymes obtained from broken chloroplasts.
The cause of this seems to be in the regenerative part of the cycle,
but the precise location of the sensitive step is yet to be discovered. This implies the same difficulty when the enzymes of this
cycle plus those that lead to starch are placed in a single reactor.
It is quite obvious that further work remains to be accomplished in
this area.

The single reactor scheme could employ stabilized but soluble
enzymes homogeneously scattered throughout the reactor. If the
enzymes in the chloroplast had some organization, while those in the
synthetic reactor did not, it would be probable that the synthetic
reactor would have slower rates, exclusive of changes caused by
differences in enzyme concentration.

Design of the carbon dioxide fixation process has been severely
hampered by the large gaps in knowledge of the mechanism and kinetics
of the enzymes, either soluble or immobilized. Though many of the
enzymes have been purified to a high degree, little is known about
their inhibitors. Therefore, it is extremely difficult to formulate
correct rate equations for their action on substrates.

Grouped Reactions

On the other hand, a system of multiple reactors, each reactor
containing a series of appropriately coupled reactions, is appealing
for several reasons: (1) inhibition problems would be minimal, (2)
the desired carbohydrate products such as fructose, glucose and
starch could be separated conveniently, (3) nonspecificity of the
enzymes leading to undesired sideproducts would be prevented, (4)
separation of ADP, $NADP^+$, and P_i would be simplified, (5) enzyme
immobilization and replenishing would be simplified, (6) regulation

and control of the operating factory would be simplified, and (7) trouble-shooting would be simplified.

The basis of this process is to group reactions in such a way that in a particular reactor, conversion of reactants to products is nearly complete. A standard free energy change of about -3 kcal/mole for an enzyme-promoted reaction will generally assure 98 to 99 percent conversion to product in an acceptable time span. Most reactions in the CO_2 fixation pathway have acceptable standard free energies ($\Delta G°$); those that do not can usually be coupled with an energetically very favorable reaction to yield the desired results. In particular, the last reaction in the total sequence within the reactor must have a high negative $\Delta G°$.

Summary of Major Approaches

In Figure 11 a summary is given of the major comparisons between the three alternative pathways for the synthesis of carbohydrate. The comparisons are made on the basis of the ATP and NADPH required, reactor size, the number of steps in the reaction, the cost of raw material, the number of enzymes, the complexity of the reactions and the energy required.

In Figure 12 the manufacturing costs are broken out. It is clear that the cellulose process is the most attractive for the near future. The fossil fuel process is penalised by the high cost of the fuel, and the CO_2 fixation process by the complexity of the systems required for ATP regeneration and enzyme production.

	Cellulose	Fossil Fuel	CO_2 Fixation
ATP req. per glucose formed	0	2	18
NADPH req. per glucose formed	0	0	12
Reactor size	smallest	moderate	largest
Number of steps	fewest (2)	moderate (5)	largest (19)
Cost of raw material	lowest	highest (glyc.)	low
Number of enzymes	smallest (1)	moderate (5)	largest (16)
Complexity of reaction	simplest	moderate	most complex
Energy requirement	lowest	moderate	highest

Figure 11. Comparison of Pathways

	Cellulose Process		Fossil Fuel Process		CO_2 Fixation
	Cellulose Glucose	Glucose Starch	Glycidaldehyde Glucose	Glucose Starch	CO_2 Starch
Processing	3.0	1.0	2.3	1.0	3*
Enzyme Production	0.4	1.4	1.8	1.4	10*
ATP Regeneration	0	3.4	6.0	2.6	60*
Raw Material	1.1	____	17.0	____	4
Total	4.5¢	5.8¢	27.1¢	5.0¢	77¢
Total for Process	10.3¢		32.1¢		77¢

*Extrapolated from the corresponding costs in the other two processes.

Figure 12. Manufacturing Cost Factors in ¢ Per Pound of Product

    It is possible that the CO_2 fixation process could become
attractive in the future as the limited supplies of fossil fuel are
exhausted and its price escalates. It should be emphasized that the
manufacturing costs were not offset in this study by the very considerable savings from the reduction in requirements for acreage
of agricultural land. It is recommended that further study be
carried out to complete the cost-benefit analysis in this way.

                    PATHWAYS STUDIED BUT REJECTED

                Formaldehyde-Transketolase (Figure 13)

    This pathway was considered because it is in large measure
identical to the CO_2 fixation pathway cycle. The overall reaction
is

        $6HCHO + 6ATP + 5H_2O \rightarrow G6P + 6ADP + 5P_i$

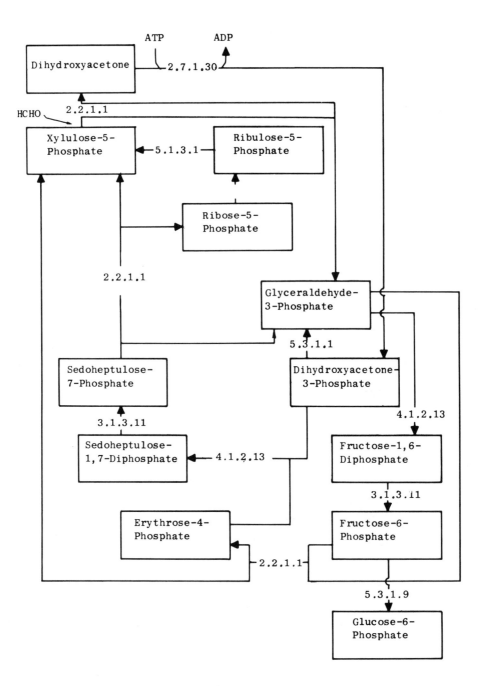

Figure 13.   The Formaldehyde-Transketolase Process for Glucose
Synthesis

This pathway was considered because of the fewer and simpler steps and lower energy requirement (6 vs 18 ATP per G6P) than CO_2 fixation. The formaldehyde can be obtained by reduction of CO_2 or from other petroleum by-products. The carboxylation, phosphoglycerate kinase, triose phosphate dehydrogenase and phosphoribulokinase reactions are eliminated, while only transketolase and triokinase reactions are gained. There is no need for NADPH.

It was concluded that this was not the best pathway for the current project for the following reasons:

(a) The transketolase reaction

$$HCHO + Xu5P \longrightarrow DHA + GALPD$$

is an unproven reaction. In an extensive literature search, no report on it was found. Since this is the main step in the entire proposed scheme, this pathway was abandoned with the recommendation that it be studied in the future. In fact, based on the data provided by Bassham and Krause, the $\Delta G°$ for this reaction has been calculated to -5.6 kcal/g-mole, a favorable one.

(b) The CO_2 reduction tends to go all the way to methanol, which has to be oxidized back to formaldehyde. The energetic economics of doing this are questionable. However, this has become only a minor reason as other hydrocarbon sources were later considered.

(c) Formaldehyde might react with F6P, S7P, and DHAP if some of the steps in the proposed scheme were to take place in the same reactor. This would complicate the reaction control problem. Also, it was feared that formaldehyde might poison some of the enzymes. However, these problems could be at least partially alleviated if separate reactors were used for the different reactions as was later considered.

Reversed Oxidative Pentose Phosphate Cycle (Figure 14)

Another possible pathway to make starch from carbon dioxide was the oxidative pentosephosphate cycle run in the reversed direction. There are fourteen steps catalyzed by twelve enzymes. The reductive pentose phosphate pathway (the selected system) was preferred over this reversed oxidative cycle for one major reason: the equilibrium of the conversion of 6-phosphogluconate to its lactone was so far to the left that the reaction could not be accomplished enzymatically and would have to be conducted at high temperatures and low pH's. Not only had lactonization of 6-phosphogluconate not been experimentally

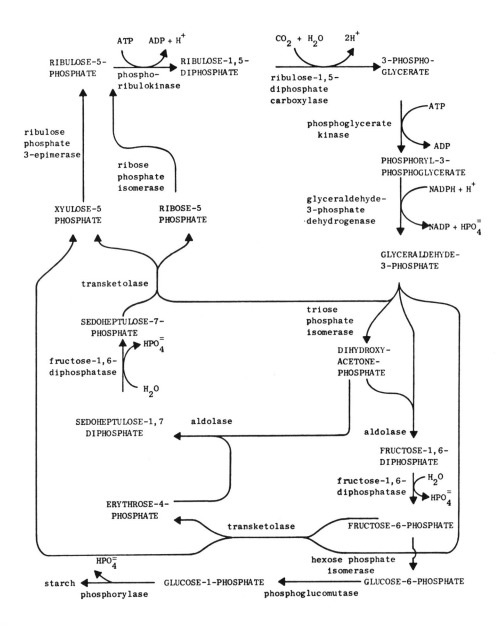

Figure 14.   Chemical Reactions-CO_2 to Starch, Reversed Oxidative
            Pentose Phosphate Cycle

demonstrated, but the isolation of 6-phosphogluconate from other phosphorylated compounds for this reaction promised to be very difficult. However, since carbamyl phosphate for the regeneration of ATP is a major portion of the cost of starch produced by the reductive pentose phosphate cycle, and since this requirement drops to 1/18 of its previous value with the reversed oxidative pathway, a second look should perhaps be taken at this process if its use were desired for special applications. For this process to be successful, however, some method would have to be found to inexpensively separate 6-phosphogluconate from the rest of the reaction mixture, and at this time no such method is readily apparent.

Location of a synthetic carbohydrate factory was studied with a view to the most promising site for providing food supplements to populations suffering calorie deficits, particularly in areas where agricultural land was scarce. India was selected.

India

Many experts do not share India's confidence that the Green Revolution will be able to sustain its growth trend. A number of reasons are cited. First, increases in crop yields have been made where they are easiest, and by the most progressive farmers who are eager to innovate and who have better land and access to water and capital. These farmers can apply fertilizer, drill wells for irrigation, and have access to transportation to move their crops to market. It is unlikely the Green Revolution will spread as rapidly among the small subsistence farmers and sharecroppers who have fewer resources unless there are fundamental changes in land tenure and credit policies.

Second, the new high-yielding seeds require irrigation; but over 75 percent of India's arable land is without assured water supplies. At the present time, the new seeds have been planted for the most part in East Punjab, where irrigated land already exists. To benefit the majority of farmers, the new grain varieties would have to spread to the remaining cropland. This would require either greater investment in irrigation projects or a breakthrough in new varieties suitable for nonirrigated agriculture.

Third, only a wheat revolution has taken place in India. The "miracle rice" has touched only a fringe of rice cultivation and has barely taken root in the rice fields of West Bengal, Oressa, and other major rice-growing areas. Since wheat accounts for 15 percent of the total acreage in food grains as against 31 percent in rice, the latter is more important in determining the overall rate of agricultural growth.

A Synthetic Carbohydrate Factory for India in the Near Future

The proposal developed for the synthetic production of carbohydrates in India concentrates on the cellulose process, named after its input material.

India is a fossil fuel importer, whereas it is the world's largest producer of bagasse, the raw material for the cellulose process. Fossil fuel imports are under strict government quotas and constitute a scarce resource valued for many purposes. Bagasse on the other hand is available in large quantities from more than 120 plants located in eight areas of the country, which include sites in Uttar Pradesh, the major source of bagasse, Bihar, East Punjab, West Bengal, Maharashtra, Andra Pradesh, Madras, and Mysore. Because of the facts outlined above, the cellulose process seems immediately preferable to a fossil fuel process in India.

An adequate technology exists in India for construction and operation of an artificial carbohydrate facility. Similar conveyers, feeders, motors, mixers, stainless steel vessels, and other necessary equipment as needed by the facility are presently being manufactured for the country's sugar and milk processing industries.

The major chemical requirement in the process is caustic soda. Since alkali products are manufactured in large quantities in India for domestic demands, it is presumed that sufficient quantities of dry flake alkali would be available for the process.

Costs and Financing

Construction costs of a complete chemical process plant in India are estimated to be 30 to 40 percent above U.S. Gulf Coast figures. These costs are on the basis of an outside contractor erecting a complete plant. They include allowances for problems in obtaining licenses and other governmental approvals, differences in material standards, import duties, delivery delays, labor costs, and many other factors which influence the cost of doing business. The cost of capital is also high. Even under the assumption that construction and operation would be entirely indigenous, it is likely that the cost of the product would be at least equal to that for U.S. production. All costs, with the exception of labor, are high.

Compatibility with National Policy

Synthetic production of food is not an alternative considered by the Indian planners in the development of their strategy to meet the

emerging food shortage crisis. Yet the projected budget outlays under
the Fourth Five Year Plan do anticipate considerable sums being
expended on agricultural research. Thus it should be possible to
find financial support for the planning of the implementation of
such a synthetic production plant. Moreover, such a suggestion would
probably constitute a significant and innovative alternative for the
Indians in their efforts to meet the challenge to produce more food
domestically.

Indian foreign trade and domestic development policies would
probably have the most important impacts on the prospective utilization of the carbohydrate production system. That policy antici-
pates that the major industries will be locally controlled, and that
further industrialization should be encouraged with governmental aid
and participation. The policies mean that any projected Indian
utilization of the system should anticipate participation by the
Indian government and that the production plant should be located
within India and be primarily if not exclusively under Indian owner-
ship and control.

Product Acceptability and Utilization

Factors influencing the acceptability of new foods are summarized in Figure 15.

---

(1)   The food must not be in conflict with cultural preferences.

(2)   It must not be associated with any taboos, religious, or
      otherwise.

(3)   It must have desirable characteristics of taste, aroma, eye
      appeal, texture, mouth feel, storage life, and flavor reten-
      tion.

(4)   Sufficient market surveying and testing must be done to insure
      that the product is in the most desirable form.

(5)   Its cost must be competitive with alternative foods.

(6)   It must be healthful if it is to fulfill the dual requisite of
      supplying energy and nutrition.

(7)   It must not be toxic when large quantities are consumed.

(8)   It must be easily prepared.

(9)   It must be properly advertised and promoted.

(10) It must not be construed to be a poor peoples' food.

Figure 15. Acceptability of New Foods

The utilization of synthetic starch in India offers two attractive possibilities: Human foods and animal feeds. For human use,
synthetic carbohydrate could be considered as part of a normal diet,
in which case it would need to be a nutritionally balanced product
in itself, or as an emergency food, where energy to sustain life
would be the prime consideration. Use of starch as a major component
in animal feeds is appealing from the standpoint of improving diet
quality by increasing available meat protein, most likely chicken.
At the same time, cereal grain for human consumption would be made
more freely available. Glucose, an intermediate product in starch
synthesis, can be used to a limited extent in either case to improve
palatability and to provide a relatively inexpensive energy source.

Human Use

Macaroni products are particularly suited to easy mass production
and can be fortified to meet nutritional needs. As dry products, they
can be easily stored and transported and offer consumer appeal for
convenience in preparation as well as permitting flavoring to suit regional tastes. Among the macaroni products, vermicelli is the only
one that has been traditionally produced in India for a long time. It
is well accepted, particularly for dessert use, which also lends a
prestige factor. This, and the success of similar products in that
region of the world, appear to indicate that pasta products would be
a reasonable choice for starch utilization as human food in India.

Animal Feeds

The utilization of synthetic starch in animal feeds is attractive
as a means of improving diet quality in a protein deficiency situation.
Chicken is the most widely acceptable meat form in India, and the
poultry industry is growing rapidly. Chickens are also an extremely
efficient convertor of feed to meat, and laying hens are efficient
biological "factories" that use feed to turn out economical eggs.
Synthetic starch as a major poultry feed component offers promise in
assisting the development of the Indian poultry industry without ab-
sorbing a disproportionate share of the food grains needed for human
use.

The efficiency possible in the poultry industry is well illustrated by a common measure--the pounds of feed required to produce
1 lb of meat. For broilers, the U.S. ratio, which was 4 in 1948, has
dropped to around 2.2. In addition, this industry can be developed
in areas where the land is not suited for other agricultural purposes.
Clearly, there is a place for large quantities of synthetic starch
and glucose in poultry feeds, particularly if they can be produced
at prices competitive with traditional feed materials.

Poultry farming being labor intensive, would create new jobs, rather than destroy existing ones, would increase the inadequate supply of protein, and would keep population in rural areas by strengthening the rural economy. Not only starch plants, but also new industries for processing and marketing poultry could be located in these areas. Furthermore, the poultry industry would help distribute the gains of economic development more equitable since the relatively low capital costs would be attractive as small-sized family operations rather than large-scaled corporate agri-businesses.

Finally, the system would protect the environment in comparison with increasing the cultivation of hybrid wheat and rice. Poultry waste is biodegradable and less dangerous than chemical fertilizers, herbicides, and pesticides needed for grain production. Equally important, it would release land in India, already in short supply. Far less acreage would be needed for an equal amount of protein produced on a poultry farm using inputs of synthetic starch compared with feed grains grown by conventional agriculture.

In short, a synthetic starch-poultry operation would have high social value, low social cost, and would permit an immediate response to immediate social and economic problems. Moreover, it might contribute materially to the basic Indian goal of abolishing hunger and malnutrition without the need for large scale importation of food.

REFERENCES

This paper is a condensation of a 297 page report entitled "Synthetic Carbohydrate: An Aid to Nutrition in the Future," prepared under the Stanford University - Ames Research Center Summer Faculty Fellowship Program in Engineering Systems Design (sponsors, NASA and the American Society for Engineering Education). NASA Contract NGR-05-020-409 to the School of Engineering, Stanford University, 1972.

Extensive references are given in the full report.

Copies of this report may be abtained from Dr. John Billingham, Code LT, Ames Research Center, NASA, Moffett Field, California 94305.

BIOLOGICAL TECHNOLOGY--PLEA FOR A NEW COMMITMENT

George T. Tsao

National Science Foundation

Washington, D.C. 20550

The 1940's and the 50's are said to be the age of nuclear physics; the 60's the age of space. The 1970's and 80's and beyond may very well be the age of biochemistry and molecular biology. That is the age when medical scientists will overcome the last few deadly diseases such as cancer and heart failure. That is the age when biochemists will learn to manipulate genetic engineering to produce food bearing, fast growing plants and meat animals. That is the age when highly nutritious and tasty synthetic food will be made from extracts of leaves and wastes. That is the age when enzymes will be broadly used as catalysts for chemical synthesis that creates no pollutants and wastes no energy. That is the age when solar energy can be absorbed via biochemical conversion to produce hydrogen and liquid fuel. BIOLOGICAL TECHNOLOGY comprises the skill and the know-how necessary for implementing the above-mentioned for increased productivity and well being of the people of this country.

Food has been a powerful political tool. The recent grain sale to the Soviet Union is at least one of the contributing factors to the relaxation of world tension. Some have said that the recent grain sale has resulted in the general increase in food prices in the US. The larger picture of the world scene, however, points to the general catching up of the US leads by other countries in all areas. The US can no longer enjoy the kind of absolute dominance in this world like she enjoyed in the immediate post second world war years. Foreign countries will continue to buy American grains and American consumers will continue to pay more for their food. Protective foreign policy and trade regulation can help release the upward pressure on food prices and head off

overseas competition only temporarily. In the long run, the real
competitive strength of this country has to derive from the
advancement of her technology.

The recent grain shipment to USSR may have affected the food
price on the domestic market. On the other hand, this country also
has a serious problem of balance of payments, and food sales to
foreign countries will help to reduce the imbalance. Generally
speaking, the strength of US exports today lies largely in areas
of high technology such as computers and modern aircraft. The US
is blessed with a huge land mass and relatively small average
population density. Japan is not likely to become a strong
competitor of food sales on the world market. Obviously, so is
this true for many other countries. To most of the underdeveloped
countries, the pressing problem is the food supply. Foreign aid
in the form of grain shipments will be most effective in winning
friends for the US around the world. Therefore, in spite of the
recent, somewhat unpleasant experience and criticism, this country
should continue to ship food to foreign lands for solving political
as well as economic and balance of payment problems. To such
activities, advancement in biological technology as applied to food
production will provide a strong support. In fact, even with large
shipments to foreign countries, food prices on the domestic market
need not go up whatsoever, if the supply and demand is well
controlled and extra food is obtained from the expanded agricultural
acreage and new, non-agricultural production methods.

It was learned during a recent visit to the USSR that, the
Soviets are committing very heavily to technology development for
food production. A new Institute in Pushkino (about 100 miles from
Moscow) was visited. The Institute employs 590 chemists, engineers,
and technicians and their sole goal is the development of new
technology for protein food production (from oil, natural gas, and
solid wastes) by microbiological means. There is no group in the
US that is anywhere even close to their size and scope. This
Institute is currently at the stage of construction of pilot plants
and installation of computers. Given a few more years, the Soviets
will surely lead the world in this area. This may not necessarily
precipitate something like a post-sputnik crisis in USA. But, the
mere fact that Soviets are doing it will have strong appeal to
many underdeveloped but overpopulated countries.

For producing one pound of beef, enough feed material containing
about 20 pounds of protein is consumed by cattle. Can human beings
be fed directly with the animal feed material? What makes grass
non-edible is not its protein content but some other minor
constituents. Isolation of protein from grass and feed material
can be a way of producing human food. Besides nutritional
problems, there are, of course, the problems of taste and texture.

Synthetic meat can be a type of textured, fibrous protein material made from spun vegetable protein fibers. The development of such technology is totally feasible but requires considerable financial support.

Single cell protein produced by growing microbes on natural gas and solid wastes provides a new source of material for food synthesis. Microbes grow extremely fast and efficiently. Every pound of single cell protein requires only 2 pounds of natural gas; while for cattle the yield is only about 5%. Microbes double their weight once in a few hours, while chickens take days and cattle take weeks.

Biological technology can also be a means of producing liquid fuel. During plant growth, solar energy is absorbed via photosynthesis and stored as the "chemical energy" of the constituents of plants. Conversion of plant constituents into liquid fuel is a means of solar energy utilization. For every pound of corn kernels, the plant also produces 1 to 1.5 pounds of other components including cobs, leaves and stocks. The US annual production of corn is 6 billion bushels (of kernels). The total solid wastes from corn contain chemical energy which is equivalent to 18% of the annual gasoline consumption of this country. With the service of reserved acreage, US production of corn can be nearly doubled, which can provide another 35% of annual gasoline equivalence. Other crops such as wheat, rice, sorghum and so on are not yet included. The current corn acreage (66 million acres) is about 30% of the total cultivated land. The amount of solid wastes from various farm crops, of course, also varies.

Besides farms, forests are another source of material which contains "stored" solar energy. An estimate given in one research proposal stated that enough alcohol can be produced from recycled newspaper, bagasse (waste from sugar cane) and sawdusts to be the equivalent of about 25% of gasoline consumption of the US. Production of alcohol from wood involves two basic steps. Wood is first converted into sugar which is in turn converted into alcohol by fermentation. The idea is not new. During the 2nd World War, both Americans and Germans developed the process for producing wood sugar which was converted to alcohol and/or yeast protein. The old process uses acids to hydrolyze wood for sugar production. Acids being non-specific catalysts also destroy the formed sugar by repolymerization via side-reaction and by forming furfural compounds which are inhibitors of the later alcohol fermentation. The overall yield thus suffers. An obvious alternative to the acid hydrolysis is the use of enzymes. Recent advancement in enzyme technology provides new impetus for a re-examination of utilization of the huge reserve of solid, farm and forest wastes as a source of sugar, alcohol, and other chemicals:

Besides cellulose and hemicellulose, lignin is the 3rd major
component of wood (about 25% of wood weight). For every ton of
paper pulp produced, there is about a ton of lignin waste which is
one of the most troublesome pollutants. There is a class of
enzymes capable of breaking lignin molecules apart.

Medical research has been testing many delicate chemicals in
search for cures of cancer and other deadly diseases. Often the
difficulty encountered by the medical scientists is the insufficient
supply of certain chemicals for their clinical studies. The enzyme,
L-asparagenase, in proposed cancer therapy provides the classical
example. A recent example is chenodeoxycholic acid, which is useful
in non-surgical removal of some types of gallstones. Currently,
this drug is made by a 8-step complicated organic synthesis and is
available only from a small company in Wales. This drug can be
made by a 2-step enzymatic process.

The current fermentation technique for producing L-asparagenase
is so inefficient that it took whole factories just to support
clinical tests. With sufficient research, fermentation yield can
be increased by 100 fold and drastically reduce the product cost.
What can be achieved by fermentation research is best exemplified
by penicillin production. During the 2nd World War, soon after
this wonder drug was first discovered, the yield of penicillin
was about 10 units per milliliter; while today the fermentation
yield of 25,000 units per milliliter is not uncommon. In fact,
the concerted research effort during and after the war for producing
penicillin was the prime factor that placed the US at the world
leading position in fermentation technology in the 1940's and 50's.
Ever since, the research in the US in this area has sagged, other
countries have moved ahead, and Japan now leads the world in many
new developments and industrial ventures in biological technology.
Currently, the number of scientists and engineers working in this
area is lopsided at about 10 to 1 between Japan and USA.

Hydrogen can be generated from water by solar energy conversion
using isolated chloroplasts and hydrogenase. After combustion water
is formed from hydrogen. This appears to be the cleanest fuel one
can ever think of. With the new advancement in stabilization by
immobilization and by chemical modification, the economical and
engineering problems for reducing the cost of chloroplast and
hydrogenase do not appear to be unsurmountable. Given 5 to 10
more years of research, an economical process for hydrogen generation should be within reach. Besides application as a fuel,
hydrogen is, of course, also an important chemical in many
industrial applications. Once a technique for direct, biochemical
absorption of solar energy is developed, it is also useful for
synthesis of many chemical compounds such as ammonia via nitrogen
fixation and high energy enzyme cofactors which are in turn the
tool for many chemical syntheses.

Since 5 years or so ago, there has been an upsurge of activities in organ transplantation in medical practice. Some have enjoyed big success and the operation becomes routine; others, however, have faltered, notably the most publicized is the heart transplantation. The reason lies mostly in the immunological rejection by the human body of foreign substances. In a broad sense, this is really another example of biological technology sagging far below the advancement of medical skills. Biological technology research on bio-materials for transplantation and other applications should include the efforts of medical scientists, chemists, material scientists and engineers. Also, the development of new diagnostic and analytical instruments will need talents from medical, electronic, and other disciplines.

Much of the harm of pollutants in water and air is biological in nature. The term BOD which stands for biological oxygen demand is a measure of the degree of pollution. The current disposal processes for all municipal wastes and much of the industrial wastes are also biological. Because of the deep concern and everyday involvement of living conditions of every citizen, new biological technology research should be of importance. EPA seems to be concerned with primarily the immediate demonstration and application of available technology. However, research on new techniques may have much better pay-off in the long run.

The energy crisis (or dilemma) is a headline item recently. With the persistently up-swinging prices, an analagous food crisis may not be far from becoming a serious problem. Fuel is certainly one of the largest industries. Food sales are also in the range of tens of billions of dollars per year. Because of its involvement in every family and every citizen, a commitment to food related research will have, as strong, if not stronger, a public and political impact as the energy research. The same can be said about biological technology related to medical applications and drug production. Of course, the same is also true with biological technology in pollution abatement research. Biological technology also has direct involvement with fuel and chemical research in hydrogen from farm and forest wastes. If the reserved farm acreage is put to use for producing crops for fuel, there is also the added incentive of saving billions of dollars currently used in farm subsidy.

Another problem that will likely become a national crisis is the supply of metals. This country is facing strong foreign competition in several metal industries. Local, rich ores are gradually depleting. Reliance on foreign metal import is also of concern for national security. Solution mining is a useful but not yet well developed technique of metal extraction by solution leaching. This technique not only shows promise in economically meeting demand by

extraction from submarginal ores, but also appears attractive from
an environmental viewpoint. Biological technology is an integral
part of solution mining because of the involvement of bacteria in
the operation. Sulfur in ores is oxidized by bacterial cells to
eventually form sulfuric acid which does the metal leaching.
Extraction by other solutions will always be affected by the
microbiological activities at the mining sites and in leaching
dumps. A sea bed is considered the last rich source of metals of
various kinds. Biological technology is also an important element
in the future development of ocean mining in understanding the
underwater marine deposits, in developing methods of metal recovery
and in maintaining and minimizing the impact on the ecological
balance.

CONTRIBUTORS

J. Adams   Design Division, Stanford University
Stanford, California 94305

F. R. Bernath   Department of Chemical and Biochemical
Engineering, Rutgers University
New Brunswick, New Jersey 08903

J. Billingham   Biotechnology Division, National
Aeronautics and Space Administration
Ames Research Center
Moffett Field, California 94035

J. A. Boundy   Northern Regional Research Laboratory
Agricultural Research Service
U.S. Department of Agriculture
Peoria, Illinois 61604

B. Chen   School of Chemical Engineering
Purdue University
Lafayette, Indiana 47907

C. K. Colton   Department of Chemical Engineering
Massachusetts Institute of Technology
Cambridge, Massachusetts 02139

C. L. Cooney   Department of Nutrition and Food Science
Massachusetts Institute of Technology
Cambridge, Massachusetts 02139

K. J. Dahl   Eastern Regional Research Laboratory
Agricultural Research Service
U.S. Department of Agriculture
Philadelphia, Pennsylvania 19118

A. Emery   School of Chemical Engineering
Purdue University
Lafayette, Indiana 47907

C. K. Gardner   Chemistry Department, Meston Rock
Old Aberdeen AB9 2UE, Scotland

B. K. Hamilton   Department of Chemical Engineering
Massachusetts Institute of Technology
Cambridge, Massachusetts 02139

259

B. T. Hofreiter   Northern Regional Research Laboratory
         Agricultural Research Service
         U.S. Department of Agriculture
         Peoria, Illinois 61604

M. J. Kolarik   School of Chemical Engineering
         Purdue University
         Lafayette, Indiana 47907

H. C. Lim   School of Chemical Engineering
         Purdue University
         Lafayette, Indiana 47907

R. A. Messing   Corning Glass Works
         Corning, New York 14830

A. C. Olson   Western Regional Research Laboratory
         Agricultural Research Service
         U.S. Department of Agriculture
         Berkeley, California 94710

N. F. Olson   Department of Food Science
         University of Wisconsin
         Madison, Wisconsin 53706

E. K. Pye   Department of Biochemistry
         School of Medicine
         University of Pennsylvania
         Philadelphia, Pennsylvania 19174

J. H. Reynolds   Monsanto Company
         New Enterprise Division
         800 North Lindbergh Blvd.
         St. Louis, Missouri 63166

T. Richardson   Department of Food Science
         University of Wisconsin
         Madison, Wisconsin 53706

S. P. Rogovin   Northern Regional Research Laboratory
         Agricultural Research Service
         U.S. Department of Agriculture
         Peoria, Illinois 61604

J. Shapira   Biotechnology Division, National
         Aeronautics and Space Administration
         Ames Research Center
         Moffett Field, California 94035

K. L. Smiley   Northern Regional Research Laboratory
                       Agricultural Research Service
                       U.S. Department of Agriculture
                       Peoria, Illinois 61604

W. L. Stanley   Western Regional Research Laboratory
                       Agricultural Research Service
                       U.S. Department of Agriculture
                       Berkeley, California 94710

G. T. Tsao   ATA Division
                       National Science Foundation
                       Washington, D. C. 20006

W. R. Vieth   Department of Chemical and Biochemical
                       Engineering, Rutgers University
                       New Brunswick, New Jersey 08903

M. V. Wondolowski   Eastern Regional Research Laboratory
                       Agricultural Research Service
                       U.S. Department of Agriculture
                       Philadelphia, Pennsylvania 19118

J. H. Woychik   Eastern Regional Research Laboratory
                       Agricultural Research Service
                       U.S. Department of Agriculture
                       Philadelphia, Pennsylvania 19118

O. R. Zaborsky   Esso Research and Engineering Company
                       Corporate Research Laboratories
                       P. O. Box 45
                       Linden, New Jersey 07036

INDEX